Evanescent Mode Microwave Components

The Artech House Microwave Library

Evanescent Mode Microwave Components

George F. Craven and Richard F. Skedd

Artech House
Boston and London

International Standard Book Number: 0-89006-176-9
Library of Congress Catalog Card Number: 87-70485

Dedication
To Dr Anthony Storr
with thanks for his encouragement
G.F.C.

To my family
R.F.S.

Contents

Preface

Every human activity is influenced by its history and the development of waveguide techniques is no exception. With a theoretical background stretching over almost 90 years, originating in the early work of Rayleigh [1] and continuing in the more practically oriented researches in the 1930s by scientists at Bell Laboratories [2] and the Massachusetts Institute of Technology [3], waveguides have a long history by electrical engineering standards. The latter workers, who initially were unaware of Rayleigh's earlier paper, were more interested in the transmission line properties of waveguides (e.g. as a connection between a transmitter and its antenna) as a lower loss alternative to the then current techniques employing open-wire and coaxial line. Consequently, they adopted the common simplification of physics, and also of transmission line theory, and assumed the line to be infinitely long and, initially, lossless. Like Rayleigh before them, they derived the equations pertaining to the fields in the guide, the cutoff conditions below which progressive wave propagation cannot exist and additionally the theoretical loss per unit length of a practical waveguide. Subsequently, waves existing below cutoff were labeled "evanescent", defined as "fading quickly". Under these specific conditions the conclusions regarding the usefulness of the waveguide as a transmission medium were only too clear, but these conclusions are only strictly correct as long as the guide is infinitely long or, equivalently, terminates in matching impedance. With a finite length of guide we must concern ourselves with the reflected "wave" and Huxley [4], for example, explained the "paradox" of the small amounts of energy transfer which occur in a cutoff guide attenuator in terms of the reflected field. The high loss which occurs in even short lengths of cutoff guide is, of course, a result of energy reflected back to the source, but the concept of loss is not without semantic difficulties because, unless specifically stated, it is not clear whether the loss is due to dissipation or reflection. Moreover, the pragmatic nature of engineering tends to establish the belief that what is not shown to be useful over a number of years is probably useless, a view that was reflected in early approximate analyses of waveguide dissipative loss which indicated that it is infinite at cutoff. It is therefore not surprising that an application in 1967 for a patent on an "evanescent mode" filter was rejected in one major country as an "inoperable device". It is reasonable to conclude that the subject of evanescent mode behavior in finite lengths of guide with a variety of terminations might be termed a "gray" area, i.e. one in which the usual engineering certainties are somewhat blurred.

However, the below-cutoff waveguide is in many ways an ideal medium for microwave components. It does not manifest the periodic resonances which characterize other distributed media and therefore is intrinsically broad band. As in any distributed constant medium, obstacles have to be introduced into the guide in order to create a resonator, but in many cases the intrinsic capacitance of a semiconductor — a diode detector or varactor — largely fulfils this role. The natural shielding of waveguides eliminates the need for boxes which are often necessary in other component media.

Pragmatism, as mentioned earlier, created some initial prejudices, but it does have its compensations. Engineers tend to be intrigued by a novel concept, especially if it promises to lead to useful applications, and a number of practical components and subsystems — many with excellent performance — have been developed in the USA, Germany and Italy [5, 6] since our early papers. Part of the purpose of this book is to encourage others to contribute to the topic. A second purpose is to encourage system applications by describing some recent achievements. Necessarily, part of this encouragement takes the form of emphasizing the "ordinariness" of this form of resonance. In one of its earliest manifestations it was the source of mysterious breakdown in waveguides carrying high power, an occurrence leading to the name "ghost mode", which implies an extraordinary or esoteric phenomenon. The title "evanescent mode" was then coined for the resonance because it is a mode which has an imaginary wave impedance (a pure reactance) and a real propagation constant.

The teaching of evanescent modes is sometimes limited to a description of the small amounts of power transmitted to a load by an evanescent TE or TM mode. However, the neglect of this resonance, which incidentally is predictable from Foster's reactance theorem, in connection with tunnel diode amplifiers some years ago resulted frequently in uncontrollable oscillation in conventional waveguide above-cutoff circuits. The time wasted on this problem then is testimony to the importance of understanding this effect. Its significance with any active device, especially if the device exhibits negative resistance, should never be under-rated.

The work involved in the development of most of the microwave components described in this book was carried out in two distinct stages. The first part occurred at Standard Telecommunication Laboratories (STL), Harlow, Essex (1965–1973), and was a specific research program solely devoted to the topic. The second phase was carried out at Marconi Space and Defense Systems (MSDS) (now Marconi Defense Systems (MDS)) and was part of a general research program into microwave components for space applications. The number of engineers employed on the work was not at any time large and in the initial phase, which covered the development of many of the basic components, was only six, including the two authors. That the work was possible at all on a topic considered "way out" at the time was mainly due to the Managing Director, the late Mr S. B. (Jock) Marsh. We are also especially indebted to our colleagues Mr Chuck Mok and Dr G. A. Hockham for a number of original contributions, and to Mr R. R. Thomas and Mr D. W. Stopp for carrying out most of the measurements, often with incidental contributions on the way.

At a later stage the work at STL switched to applications in subsystems and Dr J. Dahele and Mr D. R. Hill made significant contributions to frequency multipliers; Mr W. Kwiatkowski and Mr T. Arthanayake made earlier advances in parametric up-converters. In the work at STL, Mr G. Dawson and Mr H. B. Wood gave essential support for the development of, respectively, space components and ground station microwave links. It is also a pleasure to record the contributions of Professor J. J. Hupert of De Paul University, Chicago. Both he and the writer's colleague Dr J. S. Heeks independently recognized the significance of the theory in the field of quantum mechanics.

The work at MSDS would not have been possible without the valuable support in management of Mr D. J. Fletcher (now Managing Director, MDS). Significant contributions included those from our immediate colleagues Messrs J. V. Threthewie, Chris Norton and Phil Sleigh. Dr John Howard gave enthusiastic support for the technique elsewhere in Marconi and more recently in the USA.

Consequently, the component development, which forms the basis of subsequent systems, rests largely on the earlier work of scarcely more than a dozen engineers. If we compare this work with orthodox technology, which is the product of literally thousands of researchers spread over several decades, it would be unreasonable not to expect some shortcomings in the scope of the work described. Thus, part of the purpose of the present text is to outline these shortcomings and to indicate areas where further research would be beneficial.

The main purpose of the book is to present a coherent account of the work to date and, consequently, to provide a sound theoretical basis for future research. A secondary purpose is to illustrate its practical application so far in subsystems and thereby to provide further encouragement for other workers.

Lastly, we would like to thank STL and MDS for their permission to publish the various papers on which this text is based. MDS were especially generous in providing facilities for the production of the many drawings required in the book. We also wish to thank the following publishers for permission to adapt and reproduce illustrations originally appearing in their publications: McGraw-Hill Book Company for Fig. 6.4; the Institution of Electrical Engineers for Figs 2.2–2.7, 4.1–4.8, 7.5, 8.6–8.9; the Institute of Electrical and Electronics Engineers for Figs 2.9, 3.3–3.5, 5.7–5.8, 6.1–6.5, 6.7; ITT Europe for Figs 8.1–8.5, 8.13; and Microwaves & RF for Fig. 5.4.

George F. Craven
Bushey Heath, Hertfordshire, England

Richard F. Skedd
Bishop's Stortford, Hertfordshire, England

References

1 Lord Rayleigh. On the passage of electric waves through tubes, *Phil. Mag.*, **43**, 125, February 1897.
2 Southworth, G. C. Hyper frequency waveguides, *Bell Syst. Tech. J.*, **15**, 284, April 1936.
3 Barrow, W. L. Transmission of electromagnetic waves in hollow tubes of metal, *Proc. IRE*, **24**, 1298, October 1936.
4 Huxley, L. G. H. Survey of the principles and practice of waveguides, p. 63, Cambridge University Press, Cambridge, 1947.
5 Schunemann, K., Knockel, R. and Begemann, G. Components for microwave integrated circuits with evanescent mode resonators, *Symposium Digest, IEEE MTT International Microwave Symposium, June 1977, San Diego, CA*, p. 377, IEEE, New York, 1977.
6 D'Ambrosio, A. Realization of a KU band uncooled parametric amplifier for spacecraft applications, *Microwave '73 Conference, Brighton, May 1973*, Microwave Exhibitions and Publishers, Sevenoaks, Kent.

Chapter 1

Basic Theory

1.1 Introduction

Early investigators considered both hollow waveguides and purely dielectric guides. However, at first no-one appears to have considered the metallic rectangular structure shown in Fig. 1.1 in which both dielectric and air-filled regions are interspersed in a periodic manner. Although the mathematical techniques were available at the time it would perhaps have been unreasonable to have considered these more complex structures at that stage. However, in more recent years configurations of a periodic nature have been used in a number of slow-wave structures. Comparatively large internal reflections exist in such a combined unit and it is customary to employ Floquet's theorem [1] to determine the propagation constant of the complete region. The following analysis is taken from an unpublished internal report in 1969 by G. A. Hockham. An analysis of the properties of a comparable structure, but with metal screws replacing the dielectric, has been published by Lewin [2].

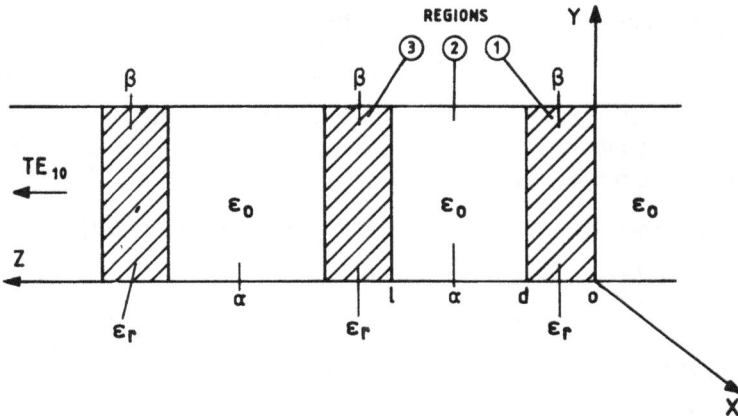

Fig. 1.1 Periodic dielectric-loaded structure.

1

1.2   Analysis of an infinite periodically loaded waveguide

Consider the propagation characteristics of the periodically loaded waveguide shown in Fig. 1.1. We require to find the propagation constant of the combined region when the waveguide is excited by a TE_{10} mode which is evanescent in the air-filled regions and propagating in the loaded regions (dielectric constant ε_r). First, expressing the fields in the two regions independently in terms of the incident and reflected waves, we have

$$E_{y_1} = \{A \exp(-j\beta z) + B \exp(j\beta z)\} \sin\left(\frac{\pi x}{a}\right) \tag{1.1}$$

Region 1

$$\frac{\partial E_{y_1}}{\partial z} = -j\beta\{A \exp(-j\beta z) - B \exp(j\beta z)\} \sin\left(\frac{\pi x}{a}\right) \tag{1.2}$$

$$E_{y_2} = \{C \exp(-\alpha z) + D \exp(\alpha z)\} \sin\left(\frac{\pi x}{a}\right) \tag{1.3}$$

Region 2

$$\frac{\partial E_{y_2}}{\partial z} = -\alpha\{C \exp(-\alpha z) - D \exp(\alpha z)\} \sin\left(\frac{\pi x}{a}\right) \tag{1.4}$$

The field in region 3 is determined by Floquet's theorem [1, 3], which states that in a periodic structure the wavefunction is multiplied by a constant when we move down the structure by a period. If, as in the present case, the structure is defined as lossless, the constant will be either purely real or purely imaginary. These conditions will define the states of attenuation and free propagation respectively. Thus, from Floquet's theorem,

$$E_{y_3}(z + l) = \exp(-\gamma l) E_{y_1}(z) \tag{1.5}$$

where γ is the propagation constant of the combined region.

Applying this theorem to the structure of Fig. 1.1 and matching the fields at $z = d$ and $z = l$ we obtain from Eqns (1.1)–(1.4)

$$A \exp(-j\beta d) + B \exp(j\beta d) - C \exp(-\alpha d) - D \exp(\alpha d) = 0 \tag{1.6}$$

$$-j\beta A \exp(-j\beta d) + j\beta B \exp(j\beta d) + \alpha C \exp(-\alpha d) - \alpha D \exp(\alpha d) = 0 \tag{1.7}$$

$$A \exp(-\gamma l) + B \exp(-\gamma l) - C \exp(-\alpha l) - D \exp(\alpha l) = 0 \tag{1.8}$$

$$-j\beta A \exp(-\gamma l) + j\beta B \exp(-\gamma l) + \alpha C \exp(-\alpha l) - \alpha D \exp(\alpha l) = 0 \tag{1.9}$$

Consequently, for a unique solution the determinant of the coefficients must be zero:

$$\begin{vmatrix} \exp(-j\beta d) & \exp(j\beta d) & -\exp(-\alpha d) & -\exp(\alpha d) \\ -j\beta \exp(-j\beta d) & j\beta \exp(j\beta d) & \alpha \exp(-\alpha d) & -\alpha \exp(\alpha d) \\ \exp(-\gamma l) & \exp(-\gamma l) & -\exp(-\alpha l) & -\exp(\alpha l) \\ -j\beta \exp(-\gamma l) & j\beta \exp(-\gamma l) & \alpha \exp(-\alpha l) & -\alpha \exp(\alpha l) \end{vmatrix} = 0 \quad (1.10)$$

Simplifying, we obtain an expression from which γ can be obtained:

$$\exp(-2\gamma l) + \frac{\exp(-\gamma l)}{\beta \alpha} [(\beta^2 - \alpha^2) \sinh\{\alpha(l-d)\} \sin(\beta d)$$

$$- 2\beta\alpha \cosh\{\alpha(l-d)\} \cos(\beta d)] + 1 = 0 \quad (1.11)$$

Let

$$M = \frac{1}{2\beta\alpha} [2\beta\alpha \cosh\{\alpha(l-d)\} \cos(\beta d) + (\alpha^2 - \beta^2) \sinh\{\alpha(l-d)\} \sin(\beta d)] \quad (1.12)$$

Then

$$\exp(-2\gamma l) - 2M \exp(-\gamma l) + 1 = 0 \quad (1.13a)$$

or

$$M = \cosh(\gamma l)$$

Thus if $|M| > 1$

$$\gamma = \Lambda \quad \text{(real)} \quad (1.13b)$$

and

$$\Lambda l = \text{arcosh } M$$

Thus the propagation constant of the region is real and represents exponential attenuation.

However, for values of $M < 1$ free propagation occurs. In particular, if we assume that $M = 0$ at the band center,

$$2\beta\alpha \cosh\{\alpha(l-d)\} \cos(\beta d) = (\beta^2 - \alpha^2) \sinh\{\alpha(l-d)\} \sin(\beta d)$$

or

$$\frac{\beta^2 - \alpha^2}{2\alpha\beta} \tan(\beta d) = \coth\{\alpha(l-d)\} \quad (1.13c)$$

where

$$\beta^2 = k_0^2 \varepsilon_r - \left(\frac{\pi}{a}\right)^2 \quad (1.14a)$$

and

$$\alpha^2 = \left(\frac{\pi}{a}\right)^2 - k_0^{\,2} \qquad (1.14b)$$

1.3   Impedance relationships

We can compare this result with the analysis of the corresponding transmission line network whose equivalent representation is shown in Figs 1.2(a) and 1.2(b) (bisected network). Expressing the parameters of Fig. 1.2(b) in chain matrix form we have

$$\begin{vmatrix} \cos\theta & jZ_{01}\sin\theta \\[2mm] \dfrac{j\sin\theta}{Z_{01}} & \cos\theta \end{vmatrix} \times \begin{vmatrix} \cosh\left(\dfrac{\alpha l}{2}\right) & jZ_{02}\sinh\left(\dfrac{\alpha l}{2}\right) \\[3mm] \dfrac{1}{jZ_{02}}\sinh\left(\dfrac{\alpha l}{2}\right) & \cosh\left(\dfrac{\alpha l}{2}\right) \end{vmatrix} = \begin{vmatrix} A_{11} & A_{12} \\[2mm] A_{21} & A_{22} \end{vmatrix} \qquad (1.15)$$

We can obtain the propagation constant (the image transfer constant) and the image impedance of the network from these matrices. The transfer constant $\cosh^2(\phi/2)$ is given by the product $A_{11}A_{22}$. Thus $\cosh\phi$ is given by

$$\cosh\phi = \left\{ \cos(2\theta)\cosh(\alpha l) + \frac{1}{2}\left(\frac{Z_{01}}{Z_{02}} - \frac{Z_{02}}{Z_{01}}\right)\sin(2\theta)\sinh(\alpha l)\right\} \qquad (1.16)$$

and for $\cosh\phi = 0$ (at midband) we obtain, after a little simplification,

$$\left(\frac{Z_{02}}{Z_{01}} - \frac{Z_{01}}{Z_{02}}\right)\tan(2\theta) = 2\coth(\alpha l) \qquad (1.17)$$

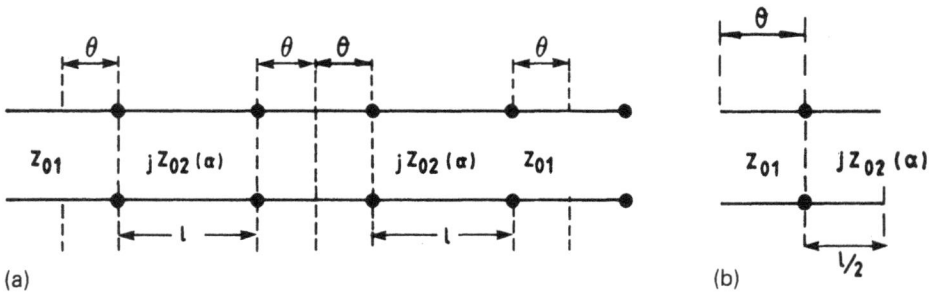

Fig. 1.2   (a) Equivalent transmission line network of a periodic structure; (b) bisected network.

Since $2\theta = \beta d$, it is evident that eqns (1.17) and (1.13c) are equal if the absolute values of Z_{01} and Z_{02} in the two waveguide sections are

$$|Z_{01}| = \frac{\omega\mu}{\beta}$$

(1.18)

$$|Z_{02}| = \frac{\omega\mu}{\alpha}$$

Thus, we can conclude that analysis using the impedance concept yields results which are identical with those of the previous analysis. In addition, we can conclude that the structure of Fig. 1.1 has a lower cutoff frequency than an unperturbed dominant-mode guide; the cutoff frequency is limited only by the magnitude of the dielectric constant in the available material. At this point it is important to distinguish this waveguide from the completely filled dielectric-loaded guide which is subject to the usual bandwidth limitations determined by higher-order mode propagation. In the present case higher-order modes would be suppressed by the evanescent sections between the dielectric. Consequently, the medium is inherently wider band than conventional dominant-mode wave-guides.

The above expression for cosh ϕ is also identical with the corresponding equation derived previously [4] for the image transfer constant of the T-section network shown in Fig. 1.3. In ref. 4 the band limit frequencies are defined more conveniently in terms of the half-lengths of the cutoff sections and are

$$f_1 = \frac{\tanh(\alpha l/2)}{2\pi Z_0 C}$$

(1.19a)

$$f_2 = \frac{\coth(\alpha l/2)}{2\pi Z_0 C}$$

(1.19b)

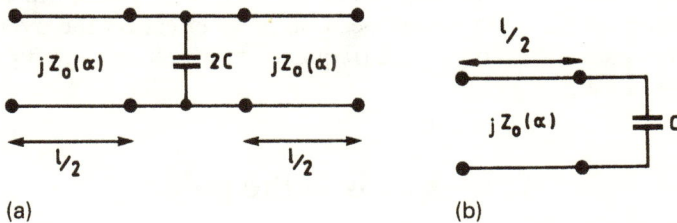

Fig. 1.3 (a) T-section resonator; (b) bisected T-section resonator. (Reprinted with permission of the *Microwave Journal*, from the August 1970 issue. © 1970 Horizon House–Microwave Inc.)

where f_1 is the lower band limit and f_2 is the upper band limit. If we define the band center in a slightly different way as the geometric mean of the band limit frequencies then

$$f_0 = \frac{1}{2\pi Z_0 C}$$

This simple expression illustrates the conditions for resonance, but it must be remembered that Z_0 is a frequency variable with its frequency dependence given by

$$Z_0 \propto \frac{1}{\{(f_c/f_0)^2 - 1\}^{1/2}}$$

In these circumstances the resonant condition is best expressed for a half-section in the form

$$Z_0 B = 1 \qquad\qquad (1.20)$$

1.4   Characteristics of evanescent mode resonators

Because transmission of power through the below-cutoff section is primarily the result of the multiple reflections between this section and the obstacle (which may take a number of forms) which stores electric energy, the individual sections comprising the infinite line may be properly regarded as resonators. Experimental resonators using dielectric obstacles have been built [5, 6]. A great deal of information concerning the resonator properties can be deduced from the two equations which define the band limits, i.e. eqns (1.19a) and (1.19b). Clearly, as the value of αl increases $\tanh(\alpha l)$ and $\coth(\alpha l)$ approach unity and the bandwidth is reduced, approaching zero in the limit as αl becomes large. It is the product αl which is important; consequently, from this viewpoint any designated waveguide size which effectively makes the operating frequency below cutoff can be used in principle. The choice of waveguide, which in general should be rectangular, is important because of its effect on the overall dimensions and dissipative loss and on the frequency at which unwanted propagation occurs above cutoff.

1.5   Fields in the guide

The field in the guide is of considerable interest. Clearly, from eqns (1.1)–(1.4), under all conditions the transverse component retains the characteristic half-sine form of the TE_{10} mode. In the z direction the component requires more detailed consideration under the special conditions imposed by (1.13c). Since the z field is the result of the interference pattern between the reflected waves in the various regions, we need to know the relationship between the constants A, B, C and D and their respective reflection coefficients. Thus, from eqns (1.6)–(1.9),

$$\frac{B}{k_1} = -\frac{C}{k_2} = \frac{D}{k_3} = -\frac{A}{k_4} \tag{1.21}$$

Consequently we can substitute

$$B = -\frac{k_1 A}{k_4} \qquad D = -\frac{k_3 C}{k_2} \tag{1.22}$$

where

$$k_1 = \begin{vmatrix} -\exp(-\alpha d) & -\exp(\alpha d) & \exp(-j\beta d) \\ \alpha \exp(-\alpha d) & -\alpha \exp(\alpha d) & -j\beta \exp(-j\beta d) \\ -\exp(-\alpha l) & -\exp(\alpha l) & \exp(-\gamma l) \end{vmatrix} \tag{1.23}$$

$$k_2 = \begin{vmatrix} \exp(j\beta d) & -\exp(\alpha d) & \exp(-j\beta d) \\ j\beta \exp(j\beta d) & -\alpha \exp(\alpha d) & -j\beta \exp(-j\beta d) \\ \exp(-\gamma l) & -\exp(\alpha l) & \exp(-\gamma l) \end{vmatrix} \tag{1.24}$$

$$k_3 = \begin{vmatrix} \exp(j\beta d) & -\exp(-\alpha d) & \exp(-j\beta d) \\ j\beta \exp(j\beta d) & \alpha \exp(-\alpha d) & -j\beta \exp(-j\beta d) \\ \exp(-\gamma l) & -\exp(-\alpha l) & \exp(-\gamma l) \end{vmatrix} \tag{1.25}$$

$$k_4 = \begin{vmatrix} \exp(j\beta d) & -\exp(-\alpha d) & -\exp(\alpha d) \\ j\beta \exp(j\beta d) & \alpha \exp(-\alpha d) & -\alpha \exp(\alpha d) \\ \exp(-\gamma l) & -\exp(-\alpha l) & -\exp(\alpha l) \end{vmatrix} \tag{1.26}$$

On expanding these determinants we have

$$k_1 = 2[\alpha \exp(-\gamma l) + j\beta \exp(-j\beta d) \sinh\{\alpha(l-d)\}$$
$$- \alpha \exp(-j\beta d) \cosh\{\alpha(l-d)\}] \tag{1.27}$$

$$k_2 = 2j\{\beta \cos(\beta d) \exp(\alpha d - \gamma l) - \alpha \sin(\beta d) \exp(\alpha d - \gamma l) - \beta \exp(\alpha l)\} \tag{1.28}$$

$$k_3 = 2j\{\beta \cos(\beta d) \exp(-\alpha d - \gamma l) + \alpha \sin(\beta d) \exp(-\alpha d - \gamma l) - \beta \exp(-\alpha l)\} \tag{1.29}$$

$$k_4 = 2[\alpha \exp(-\gamma l) - j\beta \exp(j\beta d) \sinh\{\alpha(l-d)\}$$
$$- \alpha \exp(j\beta d) \cosh\{\alpha(l-d)\}] \tag{1.30}$$

If we rearrange eqn (1.1) and substitute the appropriate reflection coefficients we have

$$E_{y_1}(z) = \frac{A}{k_4}\{k_4 \exp(-j\beta z) - k_1 \exp(j\beta z)\} \sin\left(\frac{\pi x}{a}\right) \tag{1.31}$$

If we then substitute the values for k_1 and k_4 obtained from eqns (1.27) and (1.30) we have an expression from which E_{y_1} can be obtained in the range $0 < z < d$. Similarly, from eqn (1.4) we obtain

$$E_{y_2}(z) = \frac{A}{k_4}\{k_2 \exp(-\alpha z) - k_3 \exp(\alpha z)\} \sin\left(\frac{\pi x}{a}\right) \tag{1.32}$$

Equations (1.31) and (1.32) are in a form which will permit the field to be plotted over the complete region. On substituting the values of $k_1, \ldots, k_4$, (eqns (1.27)–(1.30)) obtained for the example shown in the inset of Fig. 1.4 we obtain

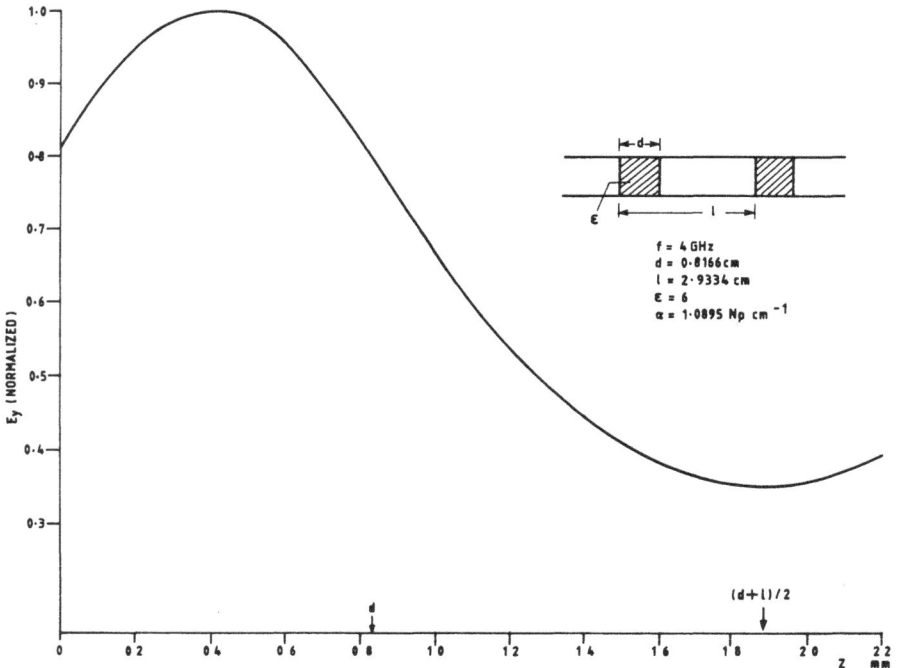

Fig. 1.4 Electric field distribution in the periodic structure at the band center.

the curve illustrating the field values which would occur. These values represent the envelope of the function, i.e. the values that would be obtained if the field were measured with a slotted line in the guide.

1.6 Q of an evanescent mode resonator

We can define the Q of a guide excited in the evanescent TE_{10} mode as the ratio of stored to dissipated energy. Assuming the conductivity of the guide walls to be large, we have [7]

$$E_y = \sin\left(\frac{\pi x}{a}\right) \exp(-\alpha z) \tag{1.33}$$

$$H_x = j\left(\frac{\varepsilon}{\mu}\right)^{1/2} \left\{\left(\frac{f_c}{f}\right)^2 - 1\right\}^{1/2} \sin\left(\frac{\pi x}{a}\right) \exp(-\alpha z) \tag{1.34}$$

$$H_z = j\left(\frac{\pi}{a}\right) \frac{1}{\omega\mu} \cos\left(\frac{\pi x}{a}\right) \exp(-\alpha z) \tag{1.35}$$

where the symbols are as defined earlier in this chapter. In more specific terms Q is defined as follows:

$$Q = \frac{\omega \times \text{peak stored energy}}{\text{power loss}}$$

$$= \frac{\omega U_P}{W_L} \tag{1.36}$$

For an evanescent TE_{10} mode

$$U_P = 2 \times \frac{1}{4} \text{Re} \int_V \mathbf{B} \cdot \mathbf{H}^* \, dV \tag{1.37}$$

where Re and the asterisk signify the real part and the complex conjugate respectively. Integrating over the volume we obtain

$$U_P = \frac{\varepsilon ab}{8\alpha} \left\{2\left(\frac{f_c}{f}\right)^2 - 1\right\} \{1 - \exp(-2\alpha l)\} \tag{1.38}$$

The power loss is given by $R_s|\mathbf{J}|^2$, where R_s is the surface resistance, $\mathbf{J} = \mathbf{n} \times \mathbf{H}$, and $\mathbf{n}$ is the unit vector perpendicular to and directed out of the wall. If there are no end walls, the loss is given by

$$W_L = \frac{1}{2} R_s \left\{ 2 \int_0^b \int_0^l |H_z|^2_{x=0} \, dy \, dz + 2 \int_0^a \int_0^l (|H_x|^2 + |H_z|^2_{y=0} \, dx \, dz \right\} \qquad (1.39)$$

$$= R_s \left(\frac{\pi}{\sigma \omega \mu} \right)^2 \left\{ a + b + \frac{a}{2} \left(\frac{f}{f_c} \right)^2 \right\} \frac{1 - \exp(-2\alpha l)}{2\alpha} \qquad (1.40)$$

Therefore the Q factor is given by

$$Q = \frac{\omega \mu}{R_s} \frac{ab}{2} \frac{1 - \frac{1}{2}(f/f_c)^2}{a\{1 - \frac{1}{2}(f/f_c)^2\} + b} \qquad (1.41)$$

The curve in Fig. 1.5 illustrates the Q of the guide as a function of frequency. It needs to be emphasized that, in general, this curve represents the Q of an inductive element; in practice it will be necessary to add the loss in the

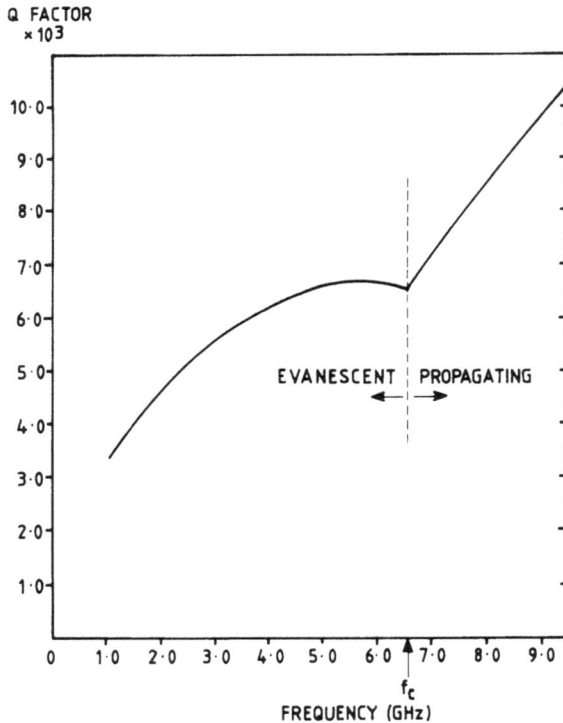

Fig. 1.5 Q factor of the TE_{10} mode in a WR90 guide.

capacitive element in order to obtain the total Q factor of the resonator. Total Q factors are commonly two-thirds of the inductive element on its own, but this requires some qualification. As cutoff is approached the "inductive" element approaches self-resonance and the amount of required additional capacitance — and therefore loss in that element — decreases. For example, resonators have been experimentally examined at 95% of the nominal cutoff frequency. At this frequency the necessary screw penetration is approximately that needed in a tuning screw in a conventional above-cutoff cavity. In these circumstances the additional loss resulting from the capacitive element is small, and the curve in Fig. 1.5 is nearly the same as the Q factor of the complete resonator.

References

1 Slater, J. C. *Microwave Electronics*, p. 170, Van Nostrand, Princeton, NJ, 1954.
2 Lewin, L. Interpretation of slimguide operation in terms of propagating (non-evanescent) waves, *Electron. Lett.*, **7** (18), 555, 9 September 1971.
3 Whittaker, E. T. and Watson, G. N. *A Course of Modern Analysis*, Section 19.4, Cambridge University Press, Cambridge, 1958.
4 Craven, G. Waveguide below cutoff: a new type of microwave integrated circuit, *Microwave J.*, **13**, 51, August 1970.
5 Amman, E. O. and Morris, R. J. Tunable dielectric-loaded microwave cavities capable of high-Q and high filling factor, *IEEE Trans. Microwave Theory Tech.*, **MTT-11**, 528, November 1963.
6 Hupert, J. J. and Vigil, J. Evanescent mode resonator of pure TE mode, *Electron. Lett.*, **4** (25), 569, 13 December 1968.
7 Craven, G. F. and Mok, C. K. The design of evanescent mode filters for a prescribed insertion loss characteristic, *IEEE Trans. Microwave Theory Tech.*, **MTT-19** (3), 307, Appendix 2, March 1971.

Chapter 2

Waveguide Obstacles

2.1 Introduction

When an obstacle is introduced into a waveguide the original unperturbed field in its immediate vicinity is distorted. The resultant field can then be resolved into the incident wave, a pair of scattered waves directed towards the source and load respectively and a series of evanescent modes. If, for the moment, we assume the obstacle to be very thin, the storage field in these evanescent modes causes it to behave as a shunt susceptance or reactance. Then, depending on whether the energy stored in the electric field exceeds or is less than that stored in the magnetic field, the obstacle is a capacitive or inductive susceptance. This commonplace proposition can be used as the basis for exploring the properties of several important obstacles in a waveguide operating below its cutoff frequency.

2.2 The "capacitive" iris

The iris under consideration, which is shown in Fig. 2.1, is usually known as the capacitive iris. Mok [1] has shown that when a TE_{10} mode is incident on such an obstacle the modes excited are neither pure TE nor pure TM; they contain both H_z and E_z components but no E_x component. The field components of these modes can be obtained from the magnetic Hertzian vector π_x (i.e. directed along the x axis) and are

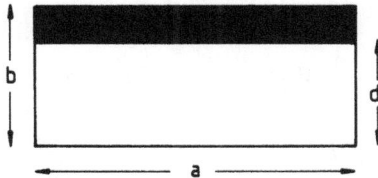

Fig. 2.1 Capacitive iris.

13

$$E_x = 0 \tag{2.1}$$

$$E_y = E_0 \sin\left(\frac{m\pi x}{a}\right) \cos\left(\frac{n\pi y}{b}\right) \exp(\mp \Gamma_{mn} z) \tag{2.2}$$

$$E_z = \mp \frac{n\pi E_0}{b\Gamma_{mn}} \sin\left(\frac{m\pi x}{a}\right) \sin\left(\frac{n\pi y}{b}\right) \exp(\mp \Gamma_{mn} z) \tag{2.3}$$

$$H_x = \pm \frac{j\{(m\pi/a)^2 - k^2\}}{k\Gamma_{mn}} Y_0 E_0 \sin\left(\frac{m\pi x}{a}\right) \cos\left(\frac{n\pi y}{b}\right) \exp(\mp \Gamma_{mn} z) \tag{2.4}$$

$$H_y = \pm \frac{j m\pi n\pi}{abk\Gamma_{mn}} Y_0 E_0 \cos\left(\frac{m\pi x}{a}\right) \sin\left(\frac{n\pi y}{b}\right) \exp(\mp \Gamma_{mn} z) \tag{2.5}$$

$$H_z = \frac{j m\pi}{ak} Y_0 E_0 \cos\left(\frac{m\pi x}{a}\right) \cos\left(\frac{n\pi y}{b}\right) \exp(\mp \Gamma_{mn} z) \tag{2.6}$$

where

$$\Gamma_{mn}^2 = \left(\frac{m\pi}{a}\right)^2 + \left(\frac{n\pi}{b}\right)^2 - k^2$$

and

$$k^2 = \omega^2 \mu\varepsilon \qquad Y_0^2 = \frac{\varepsilon}{\mu}$$

Because the TE$_{10}$ mode has a $\sin(\pi x/a)$ distribution and the obstacle has no variation in the x direction, the modes excited by it will have a $\sin(\pi x/a)$ distribution. Thus the above set of equations will apply to the capacitive iris if m is set equal to unity.

Now, as suggested in Section 2.1, the qualitative behavior of the obstacle can be determined by the difference between the time-averaged stored electric energy W_E and the corresponding magnetic energy W_M. These are given by

$$W_E = \frac{\varepsilon}{4} \operatorname{Re} \int_V EE^* \, dV \tag{2.7}$$

$$W_M = \frac{\mu}{4} \operatorname{Re} \int_V HH^* \, dV \tag{2.8}$$

where Re and the asterisk have their usual meanings. From eqns (2.1)–(2.6) the stored energy in the nth evanescent mode is

$$W_M - W_E = \frac{W\{1 - \exp(-\Gamma_{1n}l)\}}{2\Gamma_{1n}l}\left\{\left(\frac{\pi}{ak}\right)^2 - 1\right\} \qquad (2.9)$$

where $W = \varepsilon E_0^2 abl/8$ and l is the length.

Equation (2.9) shows that above the TE_{10} cutoff $1 > \pi/ak$, which leads to the well-known conclusion that in these circumstances the iris is a capacitive susceptance. However, below the TE_{10} cutoff $\pi/ak > 1$ and the sign of the susceptance changes, indicating that we now have an inductive obstacle. Clearly, the susceptance of the obstacle is zero at cutoff, i.e. the obstacle is parallel resonant. This situation is illustrated for the nth mode in Fig. 2.2. Thus this qualitative analysis illustrates the fact that, when considered over the complete frequency range, the "capacitive" iris is in fact a resonant iris with its resonance centered at cutoff. We do not have to look far in orthodox analyses to obtain confirmation of this conclusion. Lewin [2], for example, gives the expression for a capacitive iris which rearranged slightly is

$$B = \frac{4b}{\pi}\left\{\omega\varepsilon - \left(\frac{\pi}{a}\right)^2\frac{1}{\omega\mu}\right\} \ln\left\{\operatorname{cosec}\left(\frac{\pi d}{2b}\right)\right\} \qquad (2.10)$$

Comparing this with a parallel resonant circuit we have

$$C = \frac{4b\varepsilon}{\pi} \ln\left\{\operatorname{cosec}\left(\frac{\pi d}{2b}\right)\right\} \qquad (2.11)$$

and

$$L = \frac{a^2\mu}{4b\pi \ln\{\operatorname{cosec}(\pi d/2b)\}} \qquad (2.12)$$

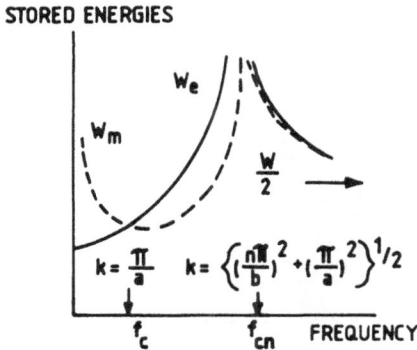

Fig. 2.2   Time-averaged stored energies of the TE_{1n} mode: W_m, magnetic; W_e, electric.

and

$$\omega_0 = \frac{\pi}{a} \frac{1}{(\mu\varepsilon)^{1/2}} \tag{2.13}$$

which is, of course, the angular cutoff frequency.

The foregoing suggests that eqn (2.10) is also valid below cutoff and will yield correct results when used in the form shown. Bearing in mind the basic nature of its derivation it would be surprising if this were not so.

2.3   Inductive iris

The inductive iris is shown in Fig. 2.3. This obstacle generates TE_{m0} modes with $E_x = 0$. The expression corresponding to (2.9) is then

$$W_\mathrm{M} - W_\mathrm{E} = \frac{W\{1 - \exp(-2\Gamma_{m0}l)\}}{kl} \left\{ \left(\frac{m\pi}{ak}\right)^2 - 1 \right\}^{1/2} \tag{2.14}$$

where

$$\Gamma_{m0}^2 = \left(\frac{m\pi}{a}\right)^2 - k^2$$

A plot of this equation in Fig. 2.4 shows that W_M exceeds W_E from zero frequency right up to the cutoff of the next propagating mode. Thus this obstacle remains inductive below cutoff. From Marcuvitz (ref. 3, Section 5.2) the susceptance of the obstacle is given by

$$B \approx -\frac{2\pi}{\omega\mu a} \cot^2\left(\frac{\pi d}{2a}\right) \left\{ 1 + \mathrm{cosec}^2\left(\frac{\pi d}{2a}\right) \right\} \tag{2.15}$$

The observation made earlier that the relevant equations concerning waveguide obstacles are also valid below cutoff deserves some expansion. At the time

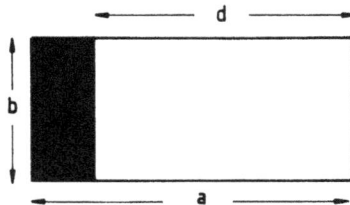

Fig. 2.3   Inductive iris.

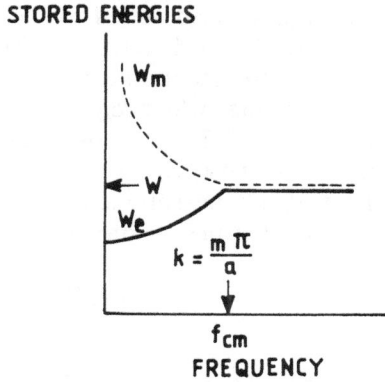

Fig. 2.4   Time-averaged stored energies of the TE_{m0} mode: W_m, magnetic; W_e, electric.

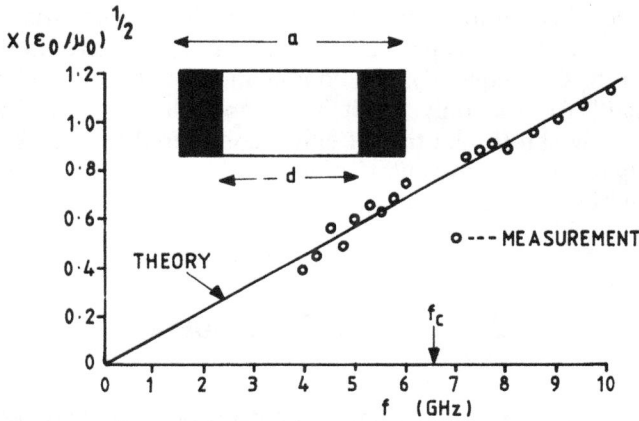

Fig. 2.5   Reactance–frequency plot of an inductive diaphragm.

of the original publication [1] no satisfactory method of measurement was available, but since then a number of methods of measuring the impedance of waveguide elements in cutoff guides have been developed and these are discussed in Chapter 9. Mok himself [4] was the first to develop a suitable technique, and the results reported in his paper include the reactance of a symmetrical diaphragm over a range including both below and above cutoff. The results are reproduced in Fig. 2.5. At that time no comparison with the theoretical results obtained from the corresponding equation

$$\frac{X}{(\mu_0/\varepsilon_0)^{1/2}} = \frac{a}{\lambda}\tan^2\left(\frac{\pi d}{2a}\right) \qquad (2.16)$$

were included, but these have now been added to the curve in Fig. 2.5. The exact agreement should be regarded with caution because the equation for a symmetrical iris is itself approximate, but it does show that in this case the statement is supported by the experimental evidence.

It will be noted that in Fig. 2.5 X is normalized with respect to the impedance of space $(\mu_0/\varepsilon_0)^{1/2}$, and the λ_g components on both sides of the equation are allowed to cancel out. Mok's measurement technique is particularly well suited to impedance measurement in the region around cutoff, and is discussed in detail in Chapter 9.

2.4   Resonant iris

The foregoing observations on the capacitive and inductive irises allow some general conclusions to be drawn concerning the resonant iris shown in Fig. 2.6. Both irises, considered separately, are effectively inductive in their properties below cutoff. Consequently, the combination of the two obstacles into a single "resonant" iris can only result in an obstacle with a higher inductive susceptance below cutoff than that of either considered singly. Therefore such an iris can only resonate above cutoff, unlike the "capacitive" iris which has been shown to have its resonance at cutoff.

2.5   Capacitive posts

The capacitive post is a very important obstacle because it can be used to realize capacitive susceptances ranging from zero to virtual infinity (at series resonance). Penetrations exceeding this value result in an inductive susceptance. This obstacle has been investigated by Mok [4] using the experimental technique

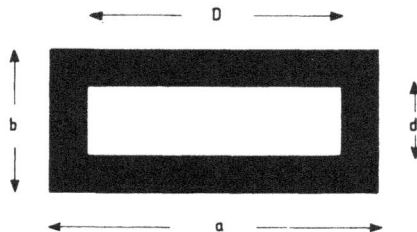

Fig. 2.6   Resonant iris.

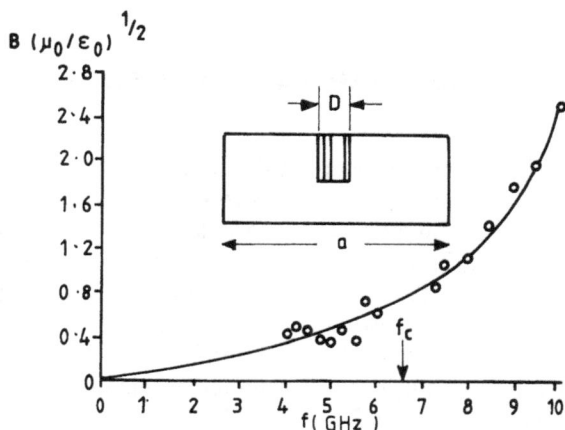

Fig. 2.7   Admittance–frequency plot of a capacitive post.

already mentioned. A curve of the susceptance, normalized with respect to free space, plotted against frequency is shown in Fig. 2.7 for an obstacle located centrally and penetrating 0.25 in (0.635 cm) into a WR90 guide. The steepening of the curve above cutoff as it approaches series resonance can be observed quite clearly. This property is a valuable feature in the control of unwanted transmission bands and is discussed in Chapter 3.

The obstacle has been investigated experimentally by Marcuvitz (ref. 3, Section 5.14) and in considerable theoretical detail by Lewin (ref. 2, Section 4.1, and ref. 5, Section 5.2) and also by Bradshaw [6]. In his first reference Lewin shows that the inductive term in the expansion for the obstacle impedance is identical with that in the expression for a thin inductive post in a waveguide. The remaining terms in the complex expression represent the effects of the capacitive element generated by coupling into the higher-order modes. Both Lewin's and Bradshaw's analyses refer to a waveguide above its cutoff frequency, and it is an open question how accurately these results apply below cutoff. Obviously, from the curve in Fig. 2.7 there is qualitative agreement. Further work is needed here to establish a quantitative relationship, and clearly some assumptions regarding the current at the end of the post will be necessary. As a tuning element in a cutoff guide the post penetrates quite deeply and consequently the current at the end of the post will not be negligible.

There is considerable theoretical interest in this topic, but the qualitative understanding already obtained is sufficient to illustrate its potential usefulness in filters. If the post consists of a screw, in practical applications it is merely necessary to adjust the screw penetration into the guide until the desired resonance is obtained. This may then form the basis for the dimensions of a post fixed to the guide wall. In this respect its problems are closely related to those of the capacitive strip.

2.6   The capacitive strip

The capacitive strip has been analyzed by Chang and Khan [7] and is illustrated in Fig. 2.8. Their approach was to investigate the regions both above and below cutoff. Their investigation into below-cutoff behavior was centered on the use of the obstacle as an alternative to the capacitive post in the filters described in this book. The method of measuring the susceptance of the obstacle was to locate it in a resonator using the equivalent circuit described by Craven and Mok [8] and discussed in Chapter 3.

Because width is an important parameter of this obstacle, a wider range of susceptance properties than can be obtained with the capacitive post is possible. Consequently, some general discussion in advance of any analytical results is desirable. First, two different resonances are possible with the obstacle considered as a separate element: a series resonance similar to that obtainable with the capacitive post and a parallel resonance which is a function of the width. The series resonance is the result of the cancelling out of the inductive property of the strip by its increased capacitance as it approaches the bottom guide wall. Greater penetration yields an inductive susceptance, and the final value — that of an inductive strip of the specified width — is reached when the strip is in contact with the bottom guide wall. Parallel resonance can occur with any penetration of the strip into the guide and is a function of the strip width. Clearly, when the width is so great that the iris is in contact with both guide walls it is an inductive iris (Section 2.2). Between these two values there must be a value at which the susceptance is zero, i.e. parallel resonance. Of course, at the resonant frequency the obstacle has no effect on the guide and represents the resonant iris below cutoff, corresponding to the above-cutoff configuration shown in Fig. 2.6. No use has been found for the obstacle in this form as yet, but wide strips with deep penetration into the guide can introduce significant loss because of the large circulating current in the obstacle near resonance.

So far we have been considering the characteristics of the obstacle *per se*, i.e. whether it is capacitive, inductive or resonant in some form. Unlike obstacles

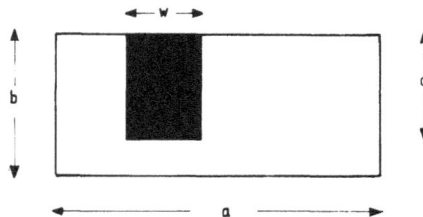

Fig. 2.8   Capacitive strip.

in an above-cutoff guide it can have all these properties. The effect that the properties have in a predominantly inductive environment must now be considered. In their analysis Chang and Khan [7] consider only narrow strips; thus the current variation across the strip can be neglected. This simplifies somewhat an already complex problem. Figure 2.9 reproduces Fig. 6 of their paper which shows the susceptance of a capacitive strip in a cutoff guide and illustrates the practical difficulties of using this obstacle as a tuning element in a resonator. Normalized susceptances in the region of 2 are necessary in order to establish resonance in a below-cutoff environment, and these are not achieved in the guide (0.9 in by 0.4 in internal dimensions) until the depth of penetration is close to the bottom wall. The penetration can be greatly reduced by a capacitive plate mounted on the end of the obstacle, but of course this would complicate the theoretical problems of analysis. Craven has employed capacitive plates as a means of increasing the gap — a necessary modification in satellite applications where high power may be involved. The problems of multipaction in these filters in a space environment are discussed in Chapter 4. The attraction of this obstacle at present lies in the fact that its susceptance can be calculated, thus eliminating the experimental work that is necessary with the capacitive post. Little improvement is likely to be achieved by increasing the width of the post, even if the analytical problems can be solved, because of the parallel resonance phenomena mentioned earlier in this section. However, the obstacle could find an application in filters constructed in square guides because the gap between the obstacle and the bottom wall is then quite large.

On all the evidence, Chang and Khan's paper presents a valuable contribution to obstacle theory in waveguides below cutoff. Above cutoff, where its

NORMALIZED |B|

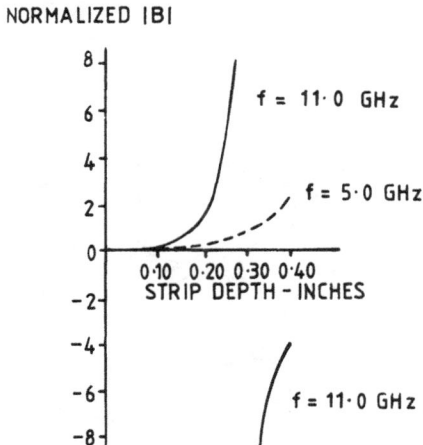

Fig. 2.9 Normalized values of the strip susceptance as a function of depth calculated for a centered strip with $w = 0.12$ in: ———, $f = 11.0$ GHz; — — —, $f = 5.0$ GHz.

contribution is also substantial, it further illustrates the increase in susceptance — characteristic of this general type of obstacle — as series resonance is approached. The value of this resonance in suppressing unwanted propagation above cutoff has already been described.

2.7　Dielectric strips

Dielectric strips (Fig. 2.10) represent the simplest obstacle analytically. If a very thin section of material with a high dielectric constant is inserted into the guide, then that section of guide will propagate and, considered independently, can be regarded as a T-section of transmission line. This was, of course, the way in which this resonance first manifested itself as the "ghost mode". If the free-space wavelength λ_0 and the guide wavelength λ_g in the dielectric are related by

$$\lambda_g = \frac{\lambda_0}{\{\varepsilon - (\lambda_0/2a)^2\}^{1/2}} \tag{2.17}$$

and the wave admittance Y_P of the propagating section is given by

$$Y_P = \left(\frac{\varepsilon_0}{\mu_0}\right)^{1/2} \frac{\lambda_0}{\lambda_g} \tag{2.18}$$

and the corresponding admittance of the cutoff section is

$$|Y_0| = \left(\frac{\varepsilon_0}{\mu_0}\right)^{1/2} \left\{\left(\frac{\lambda}{\lambda_c}\right)^2 - 1\right\}^{1/2} \tag{2.19}$$

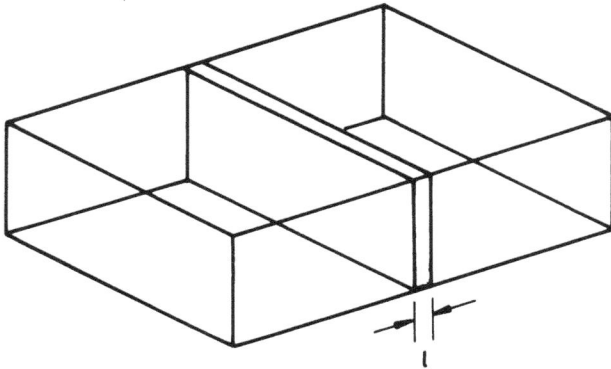

Fig. 2.10　Dielectric strip.

the susceptance of the shunt element normalized with respect to the magnitude of the cutoff guide wave admittance is

$$\frac{B}{|Y_0|} = \frac{Y_P}{|Y_0|} \sin\left(\frac{2\pi l}{\lambda_g}\right) \tag{2.20}$$

The fact that this obstacle does not excite higher-order modes makes it ideal for wider-band filters in which it is essential to restrict the excitation of these modes. The question of dissipative loss in the dielectric always arises and makes the choice of dielectric quite important. A wide range of dielectric materials with high ε and small tan δ is currently available in world markets. Since dissipative loss is inversely proportional to bandwidth, wide-band filters with acceptable loss can be constructed using materials with moderate values of tan δ. Where the lowest possible loss is essential a quartz dielectric should be employed. Its moderate constant ($\varepsilon \approx 4$) means that, if the unloaded guide is well below cutoff, the thickness required for resonance may invalidate the assumption of a thin obstacle. Allowance for the variation in eqn (2.19) across the band will then have to be included in the design, perhaps by analysis of a proposed design and subsequent iteration. The low loss properties of quartz make it suitable for use in narrow-band filters in the rather different form of a dielectric post.

2.8   Dielectric posts

The practical value of dielectric posts in narrow-band filters is considerable when the material used is high quality quartz. The loss of such filters in square guides is not substantially inferior to that of orthodox waveguide filters of the best construction. Analyses of this type of obstacle, shown in Fig. 2.11, exist

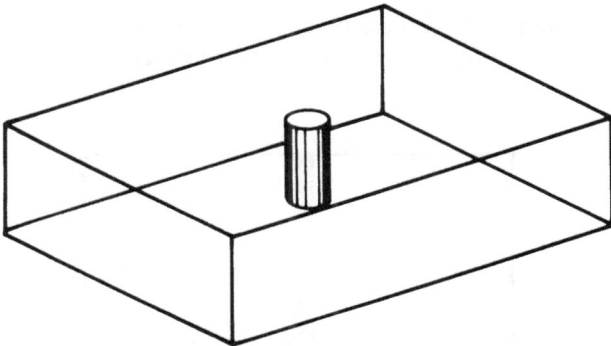

Fig. 2.11   Dielectric post.

in the literature (ref. 3, Section 5.12), but they are quite complex and their appropriateness to the problem in cutoff guides is not clear. Pending a more complete understanding of the problem, recourse to experimental measurement, wholly in the 7 GHz frequency band, has been made. These measurements are quite simple. Samples of various diameters are chosen, and the one which resonates at a slightly lower value than the chosen frequency is selected. The dielectric rod can then serve as a basis for a tunable obstacle in a similar way to the capacitive screw. A typical example is shown in Chapter 3.

2.9   Inductive post

The inductive post is illustrated in Fig. 2.12. As in the case of the inductive diaphragm or iris, the standard equations can be used to determine the susceptance of the circuit element in the guide. Their use is illustrated in the inductively loaded filters described in Chapter 4. Expressed conventionally in terms of guide wavelength, the susceptance of a single post centered in the guide is given by

$$\frac{B}{Y_0} = -\frac{2\lambda_g}{a\{\ln(a/11.6067r) + 0.2076a^2/\lambda^2\}} \tag{2.21}$$

For two posts we obtain

$$\frac{B}{Y_0} = -\frac{3\lambda_g}{a\{\ln(a/18.2157r) + 0.0771a^2/\lambda^2\}} \tag{2.22}$$

For three posts (ref. 5, Section 5.1.3, and ref. 9)

$$\frac{B}{Y_0} = -\frac{4\lambda_g}{a\{\ln(a/24.66r) + 0.0407a^2/\lambda^2\}} \tag{2.23}$$

If we wish to normalize the expression with respect to free space the guide wavelength term is removed from each side and is replaced by $(\varepsilon_0/\mu_0)^{1/2}$ on the left-

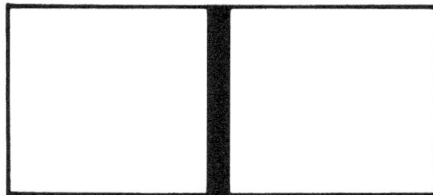

Fig. 2.12   Inductive post.

hand side of the equation and by λ on the right-hand side. However, it must be remembered that what we are normally interested in is the magnitude of this admittance/reactance relative to the corresponding value of the absolute magnitude of the wave admittance. It is this relationship which will be important in determining the resonant condition or the effect of placing an element of this type in shunt with the guide. Thus the equation to be used is

$$\frac{B}{|Y_0|} = \frac{2\lambda}{\{(\lambda/\lambda_c)^2 - 1\}^{1/2} \, a\{\ln(a/11.6067\,r) + 0.2076(a/\lambda)^2\}} \qquad (2.24)$$

with corresponding modifications to the remaining equations (2.22) and (2.23). Only the single- and double-post schemes have been used in practical applications. The post triplet finds application in an obstacle intended for use in wide-band filters.

2.10   Triplet capacitive posts

Triplet capacitive posts are related to the third example of the inductive posts, shown in Fig. 2.13. The relationship lies in the cancellation of the many higher-order TE modes that occur with the inductive obstacle. As reported by Snyder [10], these modes are the main source of the unwanted coupling that occurs with wider-band filters when the spacing between the central capacitive single posts is small. If this is the case, a post triplet will reduce this source of coupling. This is an obstacle worthy of more detailed investigation because, as mentioned in Section 2.5, the TE modes excited by such a centrally located post are identical with those generated by a corresponding inductive post. For such a scheme to be successful it is necessary that the attenuation of the TM modes, which are also excited by the capacitive post, is sufficient to prevent undesired coupling. This is unlikely to be the case in square guides but is certainly possible in rectangular guides where the technique is more likely to be required.

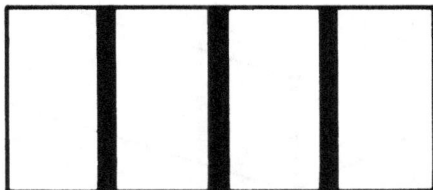

Fig. 2.13   Triple inductive posts.

2.11   Slots

One way of regarding an evanescent mode resonance is as a type of localized transverse resonance. Although this kind of reasoning is approximate because it omits the effect of the inductive environment on either side, it is a useful qualitative way of regarding the resonance. From this viewpoint we can consider the central loading provided by a capacitive post as similar to that obtained when the guide is continuously loaded by a ridge. Thus we would expect the series loading of a slot below resonance, as shown in Fig. 2.14, to produce a similar effect on this transverse resonance. This proves to be the case: the series inductance provided by this obstacle produces this result. Since the slot is open to space it will, of course, radiate and so will need to be covered by a section of waveguide which will act as a shield. However, this additional element can have its advantages. Because it is so narrow, a slot provides an ideal interface between the passive circuitry of the waveguide and a semiconductor, non-linear or active device. The series inductance introduced by leads from the semiconductor to the slot can be made very small. The additional section of guide which is necessary for shielding purposes can be employed to provide suitable reactive terminations for image frequencies generated in mixers for example. Corresponding terminations for harmonics in multipliers etc. can also be arranged.

2.12   Reflection coefficient in cutoff guides

In previous sections the behavior of obstacles has been considered in terms of their reactance or susceptance. Although this is the usual way of considering

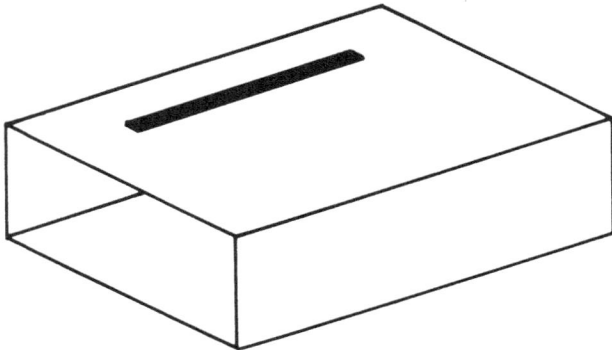

Fig. 2.14   Series inductive slot.

the properties of an obstacle, the reflection coefficient is also a relevant parameter which is often useful. The nature of the reflection coefficient in a waveguide below cutoff is critical to the transmission of energy in the guide and, in this sense, is an essential element in obtaining a conceptual grasp of the transfer of energy in such a guide. This is clearly true because, if the reflection coefficient can only have a value which is limited to the range between zero and unity, an initially decaying wave will remain substantially unchanged by the termination.

We express the voltage reflection coefficient k in the usual way as

$$k = \frac{Y_0 - Y_t}{Y_0 + Y_t} \tag{2.25}$$

If we substitute the values appropriate to a line of inductive characteristic admittance $-jY_0$, terminated in a capacitive load $Y_t = jB_t + G_t$ (with a conductance G_t in shunt), we have for the condition when the imaginary part of the termination is the conjugate of the source

$$k = -\left(1 + \frac{2jB_t}{G_t}\right) \tag{2.26}$$

Thus, if (say) $B_t/G_t = Q_t$,

$$k = -(1 + 2jQ_t) \tag{2.27}$$

It is clear from eqn (2.27) that reflection coefficients greatly in excess of unity, but imaginary (real values in excess of unity would represent gain from a passive network), occur as the condition of resonance is approached. Regarded on a transient basis the incident voltage from the inductive line is magnified by Q_t at the termination and appears as a reflected voltage. The final steady state is built up after many cycles in a similar way to that of any other resonant circuit.

2.13   Comment

It is evident from the considerations in this chapter that much of the information in the literature is generally applicable to a waveguide below its cutoff frequency. The changes in sign which occur in the reactances of some elements below cutoff are predicted by the equations derived for these elements under conventional propagating conditions. The treatments included in the early sections are principally intended to underline these changes and to facilitate a physical understanding of the reasons for them. There are also some circumstances when conclusions based on standard analyses, although qualitatively correct, yield inaccurate values. This is certainly true of the capacitive strip, as

Table 2.1 Equations for obstacle susceptance

Fig. 2.1

$$\frac{B}{|Y_0|} = -\frac{\{(\lambda/\lambda_c)^2 - 1\}^{1/2}}{\lambda} \, 8b \ln\left\{\mathrm{cosec}\left(\frac{\pi d}{2b}\right)\right\}$$

Fig. 2.3

$$\frac{B}{|Y_0|} = -\frac{\lambda}{a\{(\lambda/\lambda_c)^2 - 1\}^{1/2}} \cot^2\left(\frac{\pi d}{2a}\right)$$

Fig. 2.12

$$\frac{B}{|Y_0|} = -\frac{2\lambda}{\{(\lambda/\lambda_c)^2 - 1\}^{1/2} \, a\{\ln(a/11.6067r) + 0.2076(a/\lambda)^2\}}$$

Fig. 2.10

$$\frac{B}{|Y_0|} = \frac{Y_P}{|Y_0|} \tan\left(\frac{2\pi l}{\lambda_g}\right)$$

pointed out by Chang and Khan [7], because of the different current distribution that exists under the changed circumstances. It is likely that the qualification also applies to the capacitive post. The present view of the dielectric rod is that a new analysis of this obstacle in below-cutoff guides is probably required.

A summary of the relevant equations relating to obstacles in waveguides and their admittance properties, normalized with respect to the guide admittance, is contained in Table 2.1.

References

1 Mok, C. K. Diaphragms in evanescent waveguides, *Electron. Lett.*, **4** (3), 9 February 1968.
2 Lewin, L. *Advanced Theory of Waveguides*, p. 72, Iliffe, London, 1951.
3 Marcuvitz, N. *Waveguide Handbook*, Radiation Laboratory Series, McGraw-Hill, New York, 1948.
4 Mok, C. K. Method of obstacle admittance measurement in below-cutoff waveguides, *Electron. Lett.*, **6** (3), 50, 5 February 1970.
5 Lewin, L. *Theory of Waveguides*, Section 5.2, Newnes/Butterworths, London, 1975.
6 Bradshaw, J. A. Scattering from a round metal post and gap. *IEEE Trans. Microwave Theory Tech.*, **MTT-21** (5), 313, May 1973.

7 Chang, K. and Khan, P. J. Analysis of a narrow capacitive strip in a waveguide, *IEEE Trans. Microwave Theory Tech.*, **MTT-22** (5), 536, May 1974.
8 Craven, G. F. and Mok, C. K. The design of evanescent mode waveguide filters for a prescribed insertion loss characteristic, *IEEE Trans. Microwave Theory Tech.*, **MTT-19** (3), 295, March 1971.
9 Craven, G. and Lewin, L. Design of microwave filters with quarter-wave couplings, *Proc. Inst. Electr. Eng., Part B*, **103**, 173, March 1956.
10 Snyder, R. V. New applications of evanescent mode waveguide to filter design, *IEEE Trans. Microwave Theory Tech.*, **MTT-25** (12), 1013, December 1977.

Chapter 3

Filters

3.1 Introduction

Filters are probably the most complex of microwave components, and a suitable design theory is essential if a satisfactory amplitude–frequency characteristic is to be realized. Early evanescent mode filters were designed using the techniques of image parameter theory which, with its close analogy to transmission line theory, formed a convenient bridge between cutoff guide concepts and conventional filters in the early stages. However, this well-known approach has its disadvantages because a filter image matched at midband is badly mismatched at the band edges with consequent ripples in the transmission response.

Better results are obtained with a prescribed insertion loss theory which yields an equal ripple response across the passband. This necessitates suitable equivalent circuits to represent the elements comprising the filter. Two equivalent networks which will represent a section of waveguide below its cutoff frequency are shown in Figs 3.1 and 3.2. These networks can be deduced simply by relating the open-circuit and short-circuit impedances of the transmission line network to the corresponding T- or π-section. The resemblance to lumped-circuit elements is more than merely representational and was realized early in the investigations; it has since been commented on by a number of workers. The approximation to lumped circuits becomes closer as the frequency of operation moves further below cutoff, and virtual identity is established sufficiently far below the cutoff frequency.

The basic problem which prevents direct design in these terms arises from the fact that below cutoff the attenuation constant is a function of frequency (it is

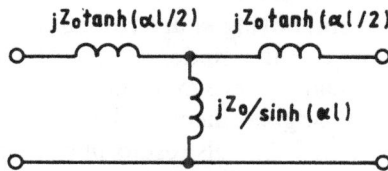

Fig. 3.1 Equivalent T-section of a cutoff guide. (Reprinted with permission of the *Microwave Journal*, from the August 1970 issue. © 1970 Horizon House — Microwave Inc.)

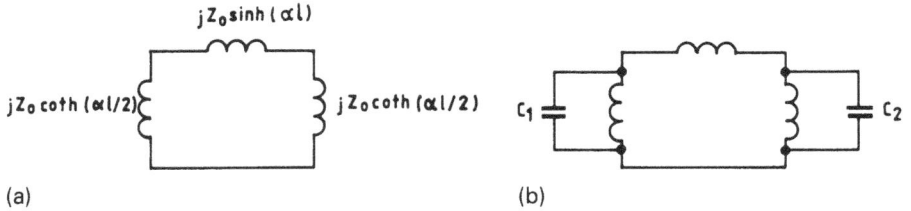

Fig. 3.2 (a) Equivalent π-section of a cutoff guide. (Reprinted with permission of the *Microwave Journal*, from the August 1970 issue. © 1970 Horizon House — Microwave Inc.) (b) Coupled-resonator filter formed by adding capacitances.

zero at cutoff). Also, the slope of the $Z_0\text{-}f$ curve becomes increasingly steep in this region, and this effect has to be taken into account in the design theory. In the original paper describing the design theory [1] a correction factor Δ was used which compensated satisfactorily in moderate- to narrow-band designs. A more refined method of correction involves the use of the frequency transformation

$$F = \frac{f}{f_0} - \left\{ \frac{(f_c/f)^2 - 1}{(f_c/f_0)^2 - 1} \right\}^{1/2} \tag{3.1}$$

The significance of this transformation may not be immediately apparent, but if we choose large values of f_c it will be evident that the expression reduces to the well-known lowpass-to-bandpass transformation

$$F = \frac{f}{f_0} - \frac{f_0}{f} \tag{3.2}$$

thus confirming the essentially lumped-circuit properties of the network.

3.2 Design theory

The foregoing transformation can be accommodated quite easily in the existing theory [1] which we now outline briefly (it is treated in greater detail in the original reference). We select the π-section representation as the more convenient one for filters embodying shunt resonators and we add capacitive obstacles at specified planes in the guide. We then have a coupled resonator filter of the type described many years ago by Shea [2].

First, returning to the original theory [1], we begin with the lowpass-to-bandpass transformation (Fig. 3.3). This results in a standard ladder network of series and shunt resonators which, for narrow-band resonators, is difficult to realize in practice. Such filters are more easily constructed in coupled-resonator form (Fig. 3.2(a)) which in any case is better suited to microwave filters. Cohn [3]

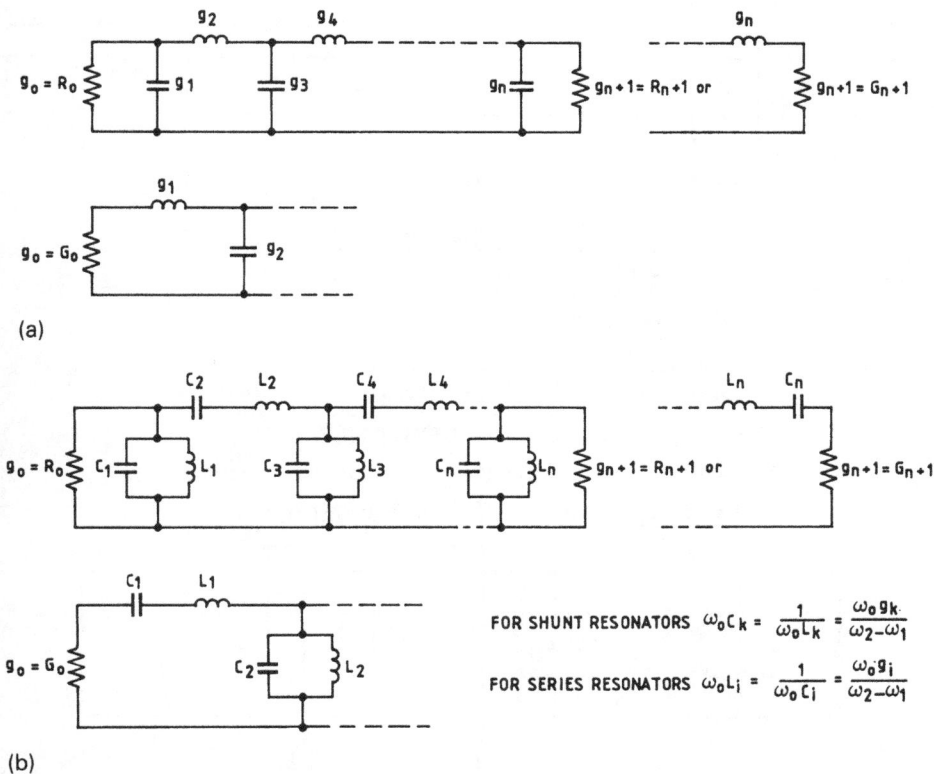

Fig. 3.3 (a) Lowpass prototype; (b) bandpass ladder network derived from a lowpass prototype.

has used the inverter concept to demonstrate the equivalence which exists between the two configurations. The network of Fig. 3.4(a) can be converted to an inverter-coupled network by means of the device illustrated in Fig. 3.4(b). It can be shown that, if we introduce negative shunt elements of equal magnitude to the series element, we have an inverter, i.e. a network with the properties of a quarter-wave line (an open circuit appears at the terminals of the inverter as a short circuit, a combination of elements in shunt appears as a corresponding combination of elements in series and so on). A characteristic of inverters of this type is that, unlike the quarter-wave line, their phase shift is invariant with frequency. However, the equivalent ladder network elements have the transformed frequency dependence of the series elements composing the inverter (Fig. 3.5). Of course, such an inverter is only possible if the combined elements in Fig. 3.4(b) add up algebraically to the original value and the remainder outside the box in Fig. 3.4(b) is positive. The changed expression for this term is shown in the figure.

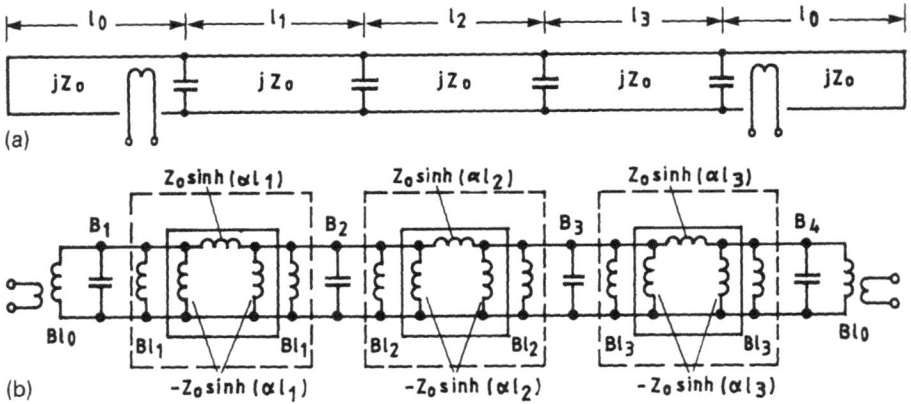

RESONANT CONDITIONS

$$B_1 = \{\coth(\alpha l_0) + \coth(\alpha l_1)\}\, Y_0 \qquad B_2 = \{\coth(\alpha l_1) + \coth(\alpha l_2)\}\, Y_0$$

$$B_3 = \{\coth(\alpha l_2) + \coth(\alpha l_3)\}\, Y_0 \qquad B_4 = \{\coth(\alpha l_3) + \coth(\alpha l_0)\}\, Y_0$$

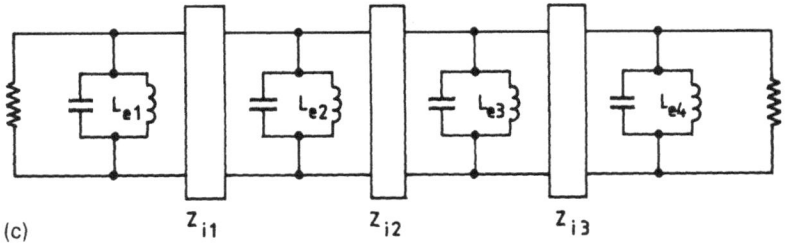

$$\frac{1}{\omega_0 L_{e,1}} = Y_0\{\coth(\alpha l_0) + \coth(\alpha l_1)\} \qquad \frac{1}{\omega_0 L_{e,2}} = Y_0\{\coth(\alpha l_1) + \coth(\alpha l_2)\}$$

$$\frac{1}{\omega_0 L_{e,3}} = Y_0\{\coth(\alpha l_2) + \coth(\alpha l_3)\} \qquad \frac{1}{\omega_0 L_{e,4}} = Y_0\{\coth(\alpha l_3) + \coth(\alpha l_0)\}$$

Fig. 3.4 (a) Transmission line equivalent circuit; (b) equivalent inverter-coupled resonators of a cutoff guide filter; (c) inverter-coupled filter and conditions for resonance.

The next stage in the design theory is to expand the input impedance of the network of Fig. 3.4(c) as a continued fraction; the values are shown in Fig. 3.5. If the ladder networks of Figs 3.4(c) and 3.3(b) are to be equal, each element of the respective networks must also be equal on a term by term basis. Thus for the first term we have

$$\frac{R}{\Delta\omega_0 L_{e,1}} = \frac{R_0 g_1 \omega_0}{\omega_2 - \omega_1} \tag{3.3}$$

where Δ takes care of the steeper reactance–frequency slope of the inductance term. Similarly, for the second term

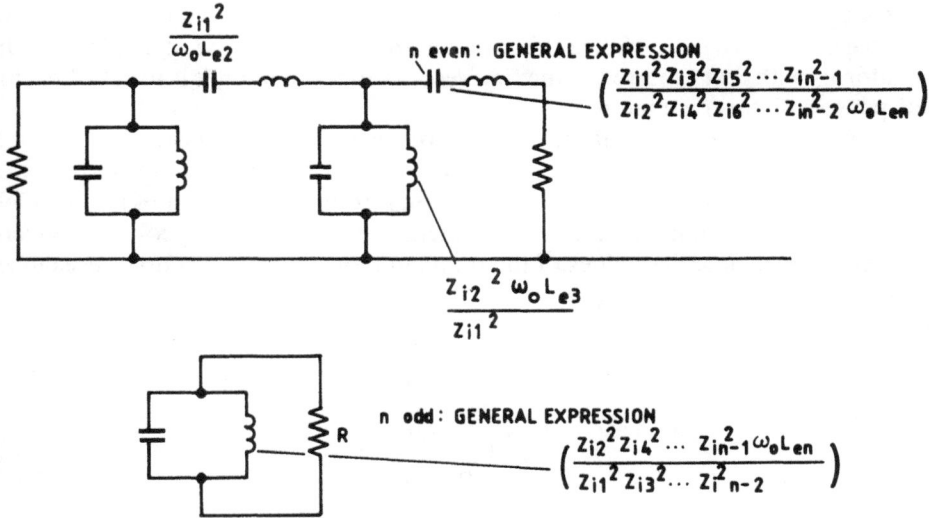

Fig. 3.5 Equivalent ladder network of the filter shown in Fig. 3.4(c) with transformed reactance values.

$$\frac{\sinh^2(\alpha l_1)}{R\Delta\omega_0 L_{e,2}} = \frac{g_2\omega_0}{R_0(\omega_2-\omega_1)} \tag{3.4}$$

Solving for $\sinh(\alpha l_1)$ and simplifying with the aid of (3.3) we obtain

$$\sinh(\alpha l_1) = \frac{\omega_0\Delta}{\omega_2-\omega_1}(g_1 g_2 \omega_0 L_{e,1}\,\omega_0 L_{e,2})^{1/2} \tag{3.5}$$

which leads to the general expression

$$\sinh(\alpha l_i) = \frac{\omega_0\Delta}{\omega_2-\omega_1}(g_i g_{i+1}\omega_0 L_{e,i}\,\omega_0 L_{e,i+1})^{1/2} \tag{3.6}$$

3.2.1 Filters beginning with a series resonator

The foregoing theory is mainly applicable to filters which are considered as separate units and are excited via coaxial connectors terminating in loops. Direct connection to a purely resistive source is possible, and the propagating guide is one of the most important examples. This is represented in the slightly idealized example of Fig. 3.6(a) which, nevertheless, forms the basis for later modifications in which the practical effects of junction susceptance, inherent in joining two guides of different dimensions, are taken into account. The essential point to be taken from Fig. 3.6(a) is that if we form a resonator by tuning the first remnant inductance (i.e. the inductance remaining after forming an inverter), then clearly the bandwidth of the first resonator will be controlled by the magnitude of the source impedance (a function of the dimensions of the propagating waveguide). Since this bandwidth is unlikely to be of a value which conforms with that required in the first resonator of the filter, it is usually best to dispense with this resonator.

As a result of this change the first resonator of the transformed circuit, which is redrawn in Fig. 3.6(b), is a series resonator. This is composed of two parts: the usual transformed element resulting from the inversion of the parallel resonator, and an additional element which, from Thévenin's theorem, is given by the equivalent series values of the remnant inductance and source resistance shown in Fig. 3.6(b). These are

$$X' = \frac{R^2 X}{R^2 + X^2} \tag{3.7}$$

$$R' = \frac{R X^2}{R^2 + X^2} \tag{3.8}$$

where $X = \tanh(\alpha l)$ and R is the source impedance.

The step which includes the junction susceptance when a practical junction between waveguides is involved is quite straightforward. If we accept that X may have the more general value that results when junction susceptance and remnant susceptance are combined in the total magnitude of X, eqns (3.7) and (3.8) can be used as they stand for the design of practical filters.

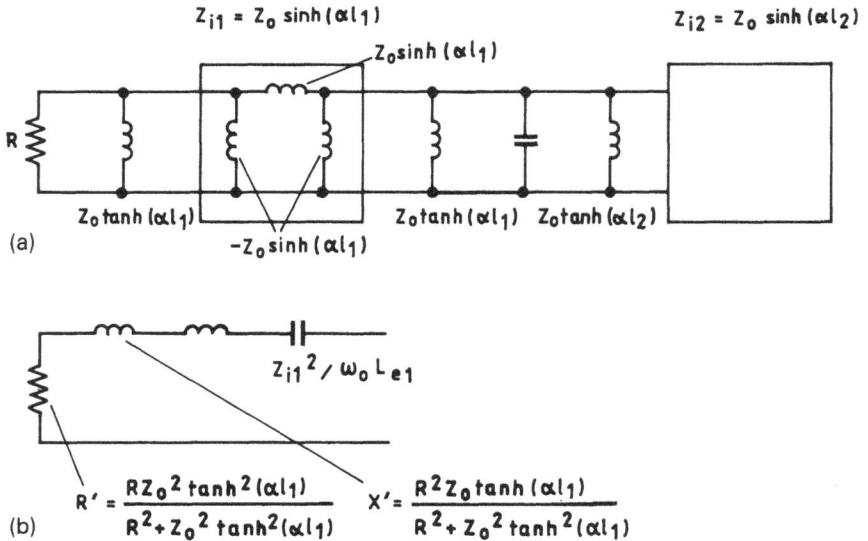

Fig. 3.6   Equivalent circuit for a filter commencing with a series resonator.

The derivation of Δ requires some explanation and is discussed in detail in the original paper [1]. Its derivation is summarized here. If we are concerned with moderate bandwidths only, the admittance of the inductive arm is given by (assuming $\alpha l \gg 1$)

$$-j Y_0 \coth(\alpha l) \approx -j Y_0 \tag{3.9}$$

$$= M \left\{ \left(\frac{\lambda}{\lambda_c} \right)^2 - 1 \right\}^{1/2} \tag{3.10}$$

where M is a constant. If the 3 dB points occur at frequencies at which the real part G and the imaginary part of the admittance are equal, then

$$-j Y_0 + j\omega C = G \tag{3.11}$$

If λ_0 is the resonant wavelength

$$\lambda = \lambda_0 (1 + \delta) \tag{3.12}$$

Substituting for λ in eqn (3.10) we can expand this expression for small values of δ. Since in the resultant expansion

$$-M \left\{ \left(\frac{\lambda_0}{\lambda_c} \right)^2 - 1 \right\}^{1/2} + \frac{2\pi C v}{\lambda_0} = 0 \tag{3.13}$$

where v is the velocity of light, we obtain the simplified expression

$$\left| \frac{\delta \lambda_0}{\lambda_0} \omega_0 C \left(\frac{\lambda_0^2}{\lambda_0^2 - \lambda_c^2} - 1 \right) \right| = G \tag{3.14}$$

For small values of δ the transmission response is symmetrical about ω_0 and the 3 dB bandwidth is given by

$$\frac{2\delta \lambda_0}{\lambda_0} = \frac{2G}{\omega_0 C [1 + 1/\{1 - (\lambda_c/\lambda_0)^2\}]} \tag{3.15}$$

Equation (3.15) can be compared with the expression for the bandwidth of a lumped circuit defined in the same way:

$$\frac{2\delta \omega_0}{\omega_0} = \frac{G}{\omega_0 C} \tag{3.16}$$

This leads to the correction factor relating the bandwidths of the two circuits:

$$\Delta = \frac{2}{1 + 1/\{1 - (\lambda_c/\lambda_0)^2\}} \tag{3.17}$$

A summary of the design equations for both types of filter is shown in Table 3.1 and is given in terms of the correction factor Δ derived above. Some readers may prefer to design the filter using the frequency transformation (3.1) mentioned earlier. This alternative requires only slight modifications to the existing theory. The essentially lowpass-to-bandpass character of the transformation is illustrated by its limiting value (3.2) expressed in the following modified form:

$$F=\frac{f'}{f_0}-\frac{f_0}{f'}$$

Thus, if we solve eqn (3.1), i.e.

$$F=\frac{f}{f_0}-\left\{\frac{(f_c/f)^2-1}{(f_c/f_0)^2-1}\right\}^{1/2}$$

using the desired band limit frequencies ($f=f_1, f=f_2$) we obtain two values for F. From these values we can find the transformed frequencies f_1' and f_2'. These frequencies are given by the two positive solutions to the quadratic

$$f=\frac{f_0 F+f_0(F^2+4)^{1/2}}{2} \tag{3.18}$$

The positive value of F yields the upper frequency f_2' and the corresponding negative value of F gives the lower frequency. The resulting expression $f_0/(f_2'-f_1')$ then replaces the corresponding expression outside the square root in the summary for these filters in Table 3.1.

3.3   Loop coupling

The design theory described in Section 3.2 applies to filters in which the load is coupled into the first resonator by a coupling loop. If the filter is either very narrow band or broad band, adjustment of the correct coupling can prove difficult in practice. In the first case this is because small changes in loop dimensions cause quite large changes in coupling and, if input match is the criterion employed, both loops need to be adjusted simultaneously. In the broad-band example the loop design necessary to achieve the desired coupling may require some attention to detail. In each case an experimental technique which enables the correct value to be definitely established is desirable. In this section we describe a suitable procedure.

The theory is more simply explained using the concept of singly loaded and doubly loaded Qs. If we consider the first resonator of a filter, e.g. a

Table 3.1

$$\Delta = \frac{2}{1 + 1/\{1 - (\lambda_c/\lambda_0)^2\}}$$

g_r, low-pass prototype element values

(a) Filters with shunt resonator terminations

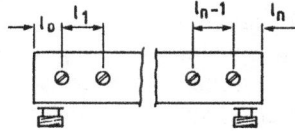

$$\sinh(\alpha l_r) = \frac{\Delta \omega_0}{\omega_2 - \omega_1} (\omega_0 L_{e,r} \, \omega_0 L_{e,r+1} g_r g_{r+1})^{1/2}$$

For maximally flat and Chebyshev amplitude response filters, $l_r = l_{n-r}$. Choose l_0 so that $\tanh(\alpha l_0) \approx 1 \approx \tanh(\alpha l_n)$. For narrow bandwidths

$$\sinh(\alpha l_r) \approx \frac{1}{2} \frac{\Delta \omega_0}{\omega_2 - \omega_1} (g_r g_{r+1})^{1/2} \text{ for } r = 1, \ldots, n-1$$

(b) Filters with series resonator

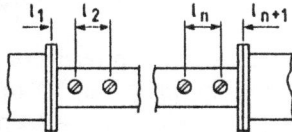

For all lengths except $r = 1$, $r = n + 1$,

$$\sinh(\alpha l_r) = \frac{\Delta \omega_0}{\omega_2 - \omega_1} (\omega_0 L_{e,r-1} \, \omega_0 L_{e,r} g_{r-1} g_r)^{1/2}$$

For maximally flat and Chebyshev amplitude response filters, $l_r = l_{n-(r-2)}$. For narrow bandwidths

$$\sinh(\alpha l_r) \approx \frac{1}{2} \frac{\Delta \omega_0}{\omega_2 - \omega_1} (g_{r-1} g_r)^{1/2}$$

The first and last lengths are equal for the above-mentioned filters, and

$$\sinh(\alpha l_1) = \left(\frac{\Delta \omega_0}{\omega_2 - \omega_1} \omega_0 L_{e,1} g_1 R' \right)^{1/2}$$

For working into propagating waveguides,

$$R' = \frac{RX^2}{R^2 + X^2}$$

where R is the transformed impedance from the propagating guide and X is the shunt reactance at the filter terminals which is equivalent to the parallel reactance of $\tanh(\alpha l_1)$ and the transformed junction reactance (see Section 3.2).

maximally flat design, the source and load resistance — as it appears at that plane — will be equal when the input loop is correctly coupled. If it is considered as an individual resonator the 3 dB bandwidth will be the value calculated from eqn (3.3). This is the doubly loaded Q of the resonator. If we now completely detune all other resonators in the filter, the first resonator will be loaded solely by the source. This is the singly loaded Q, and in the example considered where the source and load resistances are equal the singly loaded Q is obviously twice the design value. The input impedance measured at the coaxial input terminals will be a pure reactance which at resonance will be zero. (The parallel resonator is transformed to a series resonator at these terminals.) Where necessary any reference plane adjustment should be made to ensure that this condition is realized. Thus, at the frequencies where the reactance is equal to the source resistance (the singly loaded 3 dB bandwidth) the normalized input impedance on the Smith chart will be $\pm j$. If the frequencies which yield the positive and negative values of the reactance are not symmetrically disposed about the zero point, the reference plane, or the tuning, should be adjusted accordingly. With a little practice the technique provides a definite way of experimentally adjusting the coupling against an accurately measurable parameter until the correct value is obtained. Its advantage lies partly in the fact that it always indicates the direction in which improvement, i.e. decreased or increased coupling, is necessary.

The foregoing example, illustrated in terms of 3 dB bandwidths, is merely a convenient way of describing the problem. More generally, we can take any bandwidth we please. At this bandwidth the susceptance of the first resonator (or the corresponding reactance at the input terminals) must bear the same value, normalized to its load, as the corresponding bandpass prototype does to its unit load. The appropriate reactance–source resistance points on the Smith chart can be calculated from this relationship.

3.4 Approach to design programs

The previous theory was developed to meet the needs of microwave communication systems of narrow to moderate bandwidths, i.e. frequency bandwidths ranging from less than 1% to 20%. The convenience of a design theory related to the g values of the lowpass prototype is very great, and with narrowband filters the design can be accomplished in a few minutes using nothing more complex than a hand calculator. The similar ease with which the filter can be constructed makes it ideal for laboratory experiments in addition to its more sophisticated uses. However, an iterative design program is necessary for wider bandwidths. The need for this may appear self-evident from eqn (3.6): terms in al are present on both sides of the equation. However, there is a more basic reason: each inductive susceptance $1/\omega L_e$ is formed by the sum of the remnant susceptance to its left and right. For example, if we are calculating $\sinh(\alpha l_2)$, this necessitates a knowledge of $\omega L_{e,2}$ and $\omega L_{e,3}$, but in terms of normalized susceptances

$$\frac{1}{\omega L_{e,3}} = \coth(\alpha l_2) + \coth(\alpha l_3)$$

and at that stage αl_3 is not known. Thus we must begin by assuming that all αl are large, leading to $\omega L_{e,i}$ values of about 0.5, and calculate the $\sinh(\alpha l)$ values and the corresponding $\tanh(\alpha l)$ values on that basis. On the next repeat of the loop, the more realistic magnitudes of ωL_e obtained from the above sum lead to more accurate values of αl. In the range for which the theory is intended, convergence is quite rapid and a final value representing good accuracy is reached after a few iterations. However, an important exception occurs when the filter begins with an equivalent series resonator and there is a large inductive susceptance shunting the source. This limitation is fundamental to the procedure described in the lower part of Table 3.1, which is only intended for application to narrow-band filters and moderate inductive susceptances in shunt with the source. Convergence may not occur where these restrictions are infringed. Since the shunt susceptance is absorbed by reducing the coupling length, this will result in a decrease in $\sinh(\alpha l)$ and $\tanh(\alpha l)$. This decrease in the magnitude of remnant inductance will require a further decrease in $\sinh(\alpha l)$ and so on. When this happens it may be worthwhile partly to tune out some of the susceptance with a shunt capacitance. In the limit it will be preferable to add an inductive susceptance — an extra obstacle — and a tuning capacitance, and then to design the filter by including this resonator as one of the resonators in the complete chain.

3.5 Approximations in the theory

The basic approximation in the circuit theory is the assumption that the inverter is frequency invariant. This is not the case of course, as the inverter dependence is a function of $Z_0 \sinh(\alpha l)$. As mentioned earlier, at frequencies well below cutoff this dependence tends to the value ωL. Since we are interested in the factors which limit the bandwidth in the present theory, we can obtain a better understanding of its limitations by studying the expressions representing the individual elements in the equivalent ladder network in Fig. 3.5. If we accept the ωL approximation, the individual elements comprising a series resonator have an inverter dependence of ωL because the frequency component in the remaining terms of the numerator and denominator cancel out. In the shunt resonator the frequency dependence cancels out completely because the number of terms is even. The effect of this difference is to make series resonators both wider in bandwidth than the design value and, because of the imposed ωL dependence, asymmetrical about the resonant frequency in their reactance properties. The bandwidth can be corrected by a program which compares the total bandwidth of an equivalent series resonator — seen at the input terminals through all the inverters between it and these input terminals — with the corresponding prototype of the genuine ladder network. On the basis of the comparison changes are made in the inverter immediately adjacent to it, and the comparison is then repeated. Such an iterative program — which can conveniently be called a "bandwidth equalization" program — can make the bandwidths of the two networks as close to each other as desired, but it cannot completely compensate for the asymmetry which is a basic consequence of the initial assumptions.

The foregoing limitations have been emphasized because of the necessity of obtaining a complete understanding of the theory behind the filter. Nevertheless, practical filters of considerable bandwidth can be constructed using the existing theory. However, it is one of the areas mentioned earlier in which additional research is desirable.

A fundamental assumption in the theory is that only the TE_{10} mode exists in the guide. Thus higher-order modes, which are strongly excited by some obstacles, could invalidate the theory by introducing mutual coupling between the obstacles. A somewhat purist approach, which avoids these problems by employing a material of high dielectric constant as the capacitive element (Section 2.7), is practicable. Of necessity the dielectric slice completely fills the guide width, and then only the TE_{10} mode, which is excited initially, can exist in the guide. This restriction undesirably limits the range of useful obstacles, and in particular eliminates the capacitive screw which is so useful in narrow-band filters where exact tuning is essential. A screw also has the merit of cheapness. The important constraint which must be satisfied is that the attenuation constant of any higher-order mode must be sufficiently high to prevent additional coupling by these modes. If this is not done a typical overcoupled response will be obtained, i.e. a double-humped band response which is wider than expected and has a high reflection at the resonant frequency. This is always possible with bandwidths approaching 10% and is very probable when the bandwidth exceeds this value. Some techniques which mitigate the problem have been devised and are worth outlining.

The first of these is to choose the correct guide size for the filter. The ratio of the attenuation constant of the TE_{10} mode to other modes is influenced by the aspect ratio and also by the proximity of the operating frequency to cutoff. As cutoff is approached the attenuation constant of the TE_{10} mode is reduced, and consequently the length of the coupling section has to be increased to obtain the desired αl product. The attenuation constants of higher-order modes will not be proportionally decreased by moving the operating point a littler closer to cutoff, and the essential increase in coupling length (in order to maintain the design) greatly increases the overall attenuation of the higher-order modes. This reasoning runs contrary to intuitive concepts regarding wider-band filters, but for the bandwidths under consideration it has been shown to make the difference between a filter that employs capacitive screws and is viable and one that is not. A few preliminary calculations of designs with different cutoff frequencies and of the effect on the attenuation of some adjacent higher-order modes are helpful in providing the necessary corroboration in this respect. Multiple-screw configurations have been employed by Snyder [4] as a means of reducing coupling by higher-order TE modes.

The aspect ratio of the guide is also important in its effect on TM modes. Increasing the guide height, whilst helpful in raising the Q factor of the guide, decreases the attenuation constant of the TM modes. Where this must be done the use of symmetrical screws is useful in suppressing the even-order modes. The technique is analogous to that described above, but whereas a post triplet suppresses all TE modes up to and including the TE_{60}, a single capacitive post split with one half in the top wall and one half in the bottom wall suppresses only the even-order TM modes.

The concentration on capacitive screw obstacles in the early investigations introduced some controversial aspects into the work. The initial interest in this obstacle was an essential element in maintaining funding because of the promise it held for simple cheap filters. However, eventual publication [5] was met with the criticism that its premise was in error and that in effect the resonance was a TEM resonance [6] despite experimental evidence in the original letter demonstrating the resonance unambiguously in a completely different way. This information was described in more detail in a reply [7], and further work on suitable obstacles including a dielectric rod — which cannot support a TEM mode — subsequently cleared up most of the objections. However, the controversy has relevance to the foregoing sections. Some thought will indicate that the above conditions which we seek to correct in the ways described are those which normally exist in a comb-line, or interdigital, filter (i.e. when considered from a waveguide viewpoint the coupling is multimode). These are the coupling conditions in which the evanescent mode theory will not function since only one mode can be the source of the coupling. In addition, in the evanescent theory it is irrelevant whether the tuning screws are all located on one side of the guide, as in the comb-line filter, or on alternate sides, as in the interdigital filter; the design theory remains the same. Once again the similarities and differences between these filters are a demonstration of the subtleties of the microwave art; devices constructed according to one theory can look similar in one form and yet bear little or no resemblance to each other in another realization of the theory. An example is the dielectric strip already mentioned, which is essential in genuinely wide-band filters. It is also necessary in dual-mode filters of any complexity, in order to prevent uncontrolled cross-polarization coupling. These filters are discussed in a later chapter.

3.6   Temperature compensation

The use of lightweight materials such as aluminum is very attractive in applications where weight is at a premium, e.g. satellite applications. However, the coefficient of expansion of aluminum is quite large and substantial shifts in the center frequency can be expected. With a capacitive screw, which is the most probable obstacle in narrow-band filters where frequency stability will be very important, the resonant frequency depends on the gap between the capacitive screw and the bottom guide wall and also on the overall guide dimensions. This more complex relationship means that the frequency stability, computed on the basis of the temperature coefficient of the material, will not be correct for these resonators. For example, although Invar can maintain the guide dimensions relatively constant, the small expansion that remains will unduly affect the capacitive gap, where the electric energy storage is considerable, and therefore will have a disproportionate effect on the resonant frequency. Thus it is better to employ materials with different coefficients of expansion for the guide and the capacitive rod (e.g. aluminum for the guide and iron (silver-plated) for the capacitive element). The curves in Fig. 3.7(a) illustrate the degree of compensation that can be achieved with a copper guide: curve 1 represents uncompensated copper, curve 2 illustrates overcompensation with a silver-plated iron obstacle and curve 3 represents the final results obtained with a screw that is a composite of iron and copper. In determining the percentage of iron and copper required in the composite screw, it is helpful to build a cavity with a screw of each kind in the top and bottom walls respectively. Measurements of the compensation obtained with various penetrations of the screws will provide a useful, but not exact, guide to the proportions of each material in the final single screw.

The compensation of dielectric rods requires a different technique. The quartz rod is suspended from a metal bush mounted on top of the guide as shown in the inset to Fig. 3.7(b). The degree of compensation is controlled by the length of the bush, which is made from copper. Examples shown in Fig. 3.7(b) illustrate the degree of compensation, ranging from undercompensation to overcompensation, realized with different lengths of quartz.

3.7   Parasitic passbands

Above cutoff normal propagation occurs and at some frequency an unwanted passband can be expected. The frequency at which this occurs depends on a number of factors: the ratio f_0/f_c, the type of obstacle and the bandwidth of the filter. The cutoff frequency f_c is obviously of primary impor-

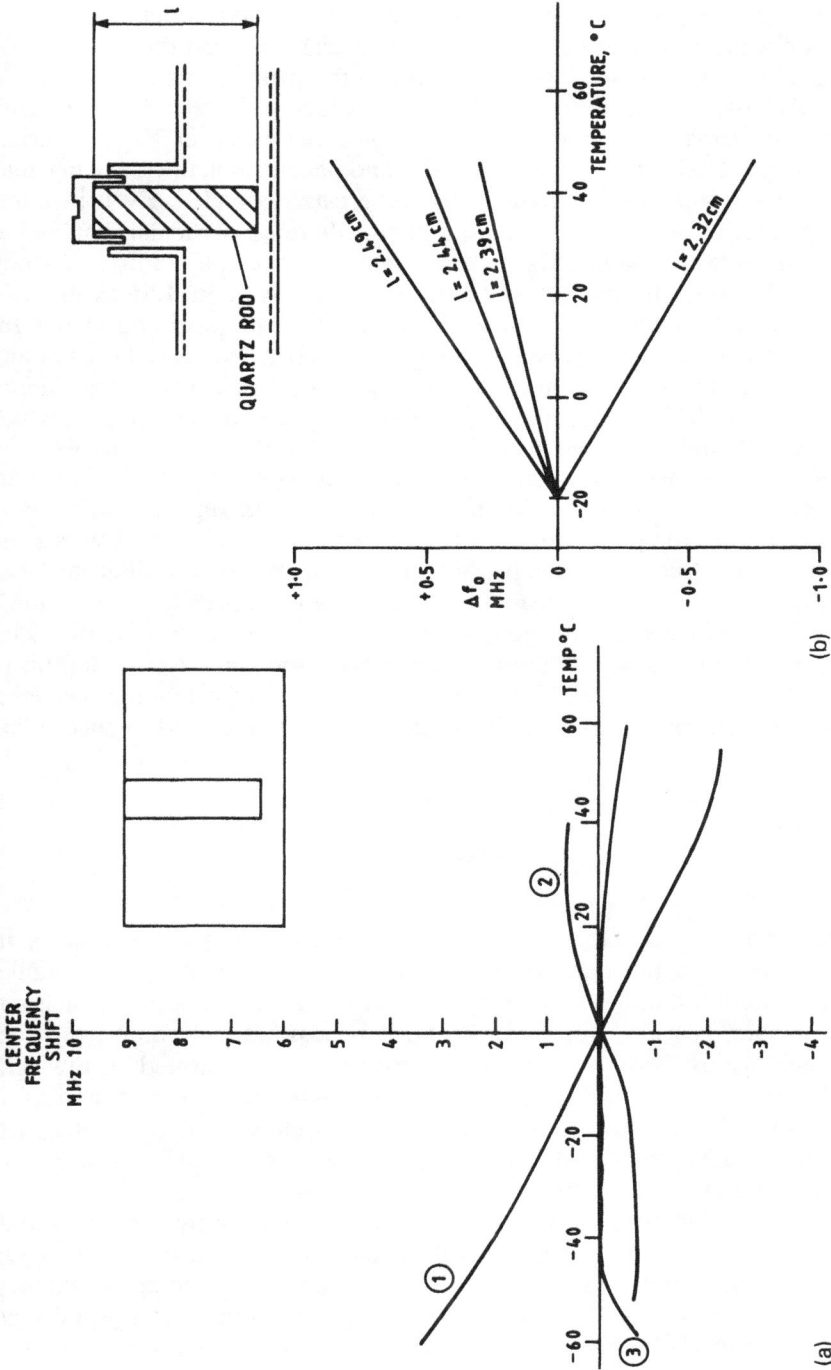

Fig. 3.7 (a) Temperature compensation of a capacitive-rod-tuned resonator; (b) temperature compensation of a quartz-tuned cavity.

tance since normal propagation can be expected above this frequency. However, this does not mean that power is transmitted straight through the unit because the capacitive obstacles are highly reflective in the propagating region. This is particularly true in the example of capacitive screws which have a series resonance at some frequency. If this resonance is located at an appropriate frequency by choosing the proper screw size and penetration, the range in which unwanted passbands are suppressed can be greatly extended. Even in filters operating fairly close to cutoff $f_c/f_0 = 1.64$ and, where no particular trouble has been taken, parasitic passbands usually occur at twice the cutoff frequency. Commercial manufacturers of these filters [8] commonly quote $4.5f_c$ as standard for the first parasitic passband, and up to 50 times the passband center for special requirements. In the presence of highly reflective obstacles this passband is primarily the result of the filter behaving as a kind of conventional direct-coupled filter, with the capacitive screws acting as the end susceptances which couple the half-guide wavelength cavities together. For this reason narrow-band filters will have parasitic passbands at lower frequencies than those in wider-band filters. This provides the clue to really broad-band suppression of these passbands. For example, if a narrow-band filter has unwanted passbands at an unacceptably low frequency, it is possible to build a broad-band filter into it as an integral unit. Because it is broader band it can be constructed in a waveguide with a high cutoff frequency without significant loss. If the capacitive obstacles then enter the guide deeply, the out-of-band rejection can be extended up to very high frequencies indeed. Examples of this method of extending the rejection band of frequencies are described later in this chapter and in succeeding chapters.

3.8   Performance

The options available for any given design create some problems in defining the performance that can be achieved with these filters because, unlike waveguides above cutoff, there is no classified waveguide size for a given frequency band. A wide choice exists, and it is possible to trade off physical dimensions against loss and mass; temperature-compensated lightweight material represents an additional alternative. Under these circumstances it seems reasonable to illustrate the performance obtainable in some standard filters in which the specifications are not severe and to supplement these results with some more extreme examples.

Figure 3.8 illustrates the performance of a comparatively simple filter. A four-resonator filter with coaxial input terminations designed for a maximally flat response had a midband loss of 0.2 dB. The filter measured approximately 15.25 cm × 5.1 cm × 2.54 cm. A second filter (Fig. 3.9) of more complex design had the dimensions 40 cm × 2.54 cm × 1.3 cm and was constructed in alumi-

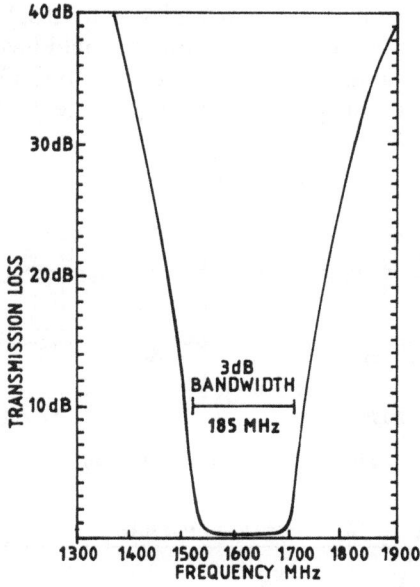

Fig. 3.8 Performance of a 1600 MHz filter.

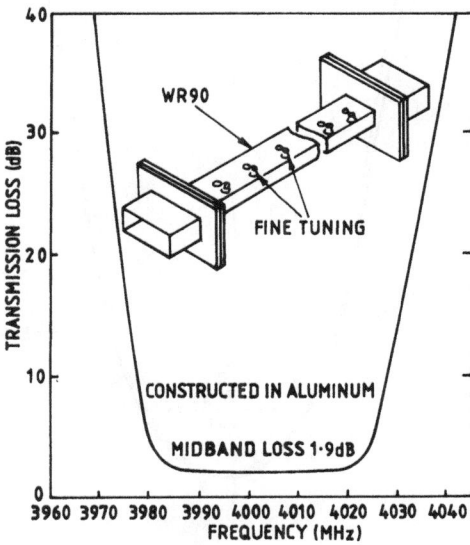

Fig. 3.9 Transmission response of a ten-cavity maximally flat filter.

num. Its narrower bandwidth resulted in a loss of 1.9 dB, but silver-plating techniques developed later for use in satellite filters would have reduced this loss to about 1.2 dB whilst retaining a low mass of only 170 g. The filter was designed to operate directly into the propagating guide; the first (and last) coupling

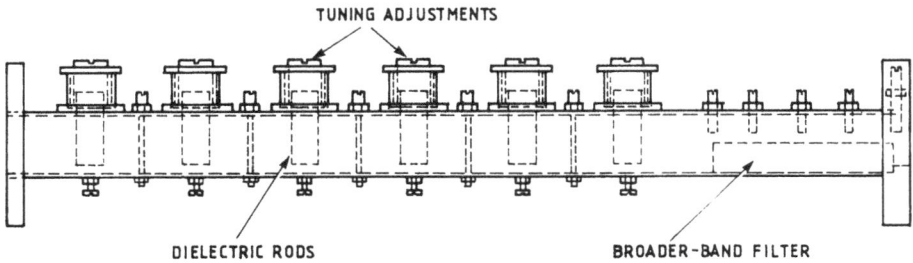

Fig. 3.10   Narrow-band filter.

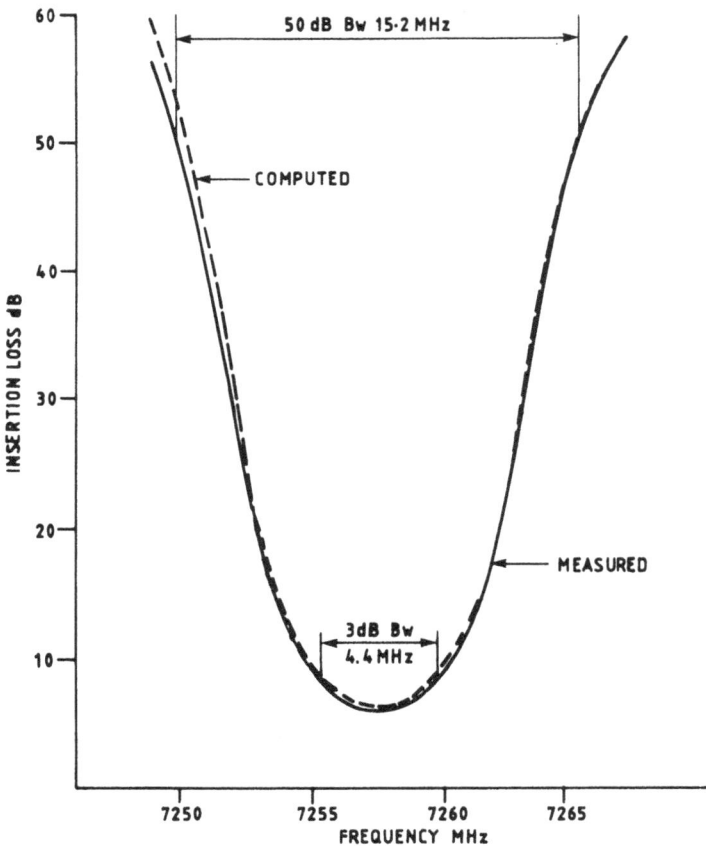

Fig. 3.11   Performance of a narrow-band filter.

lengths were modified in order to accommodate the junction susceptance resulting from the junction between the two guide widths. Practical designs of the foregoing types can often be made into a more compact and convenient form by folding the filter back on itself.

Figures 3.10 and 3.11 illustrate the technique taken to near its limits: a satellite beacon filter with a bandwidth of only 4.4 MHz centered at 7.257 GHz. Only a small bandwidth was available for the beacon signal, which could tolerate significant loss. By locating the operating frequency fairly close to cutoff (9.49 GHz), an unloaded Q of 5600 could be obtained with a satisfactory degree of temperature compensation. The midband loss as shown when scaled to the more common bandwidth of 1% would represent a dissipative loss of only 0.5 dB. Maximum unloaded Qs of about 6000 can be achieved at these frequencies, i.e. about 75% of that obtainable in a conventional waveguide cavity. The unwanted passband performance of such filters would normally be quite poor, but the model shown incorporated a broad-band rectangular guide filter (shown on the right-hand side of Fig. 3.10), with comparatively deep penetration of the screws, which provided the necessary improvement in this characteristic. This

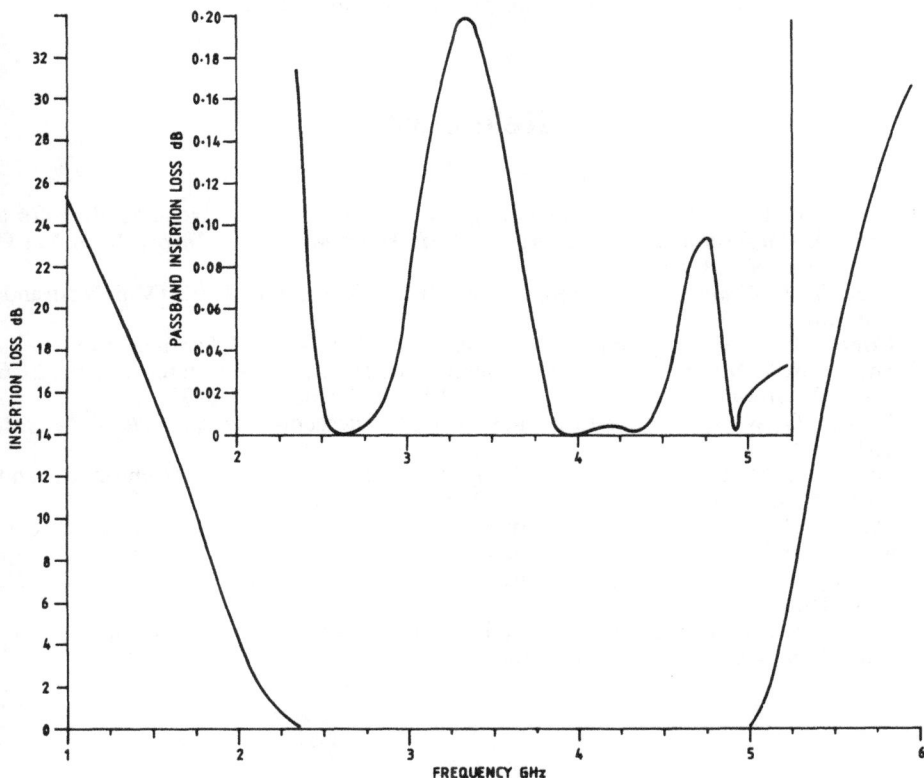

Fig. 3.12 Computed performance of a five-resonator wide-band filter.

filter included inductive loading which effectively reduced the length of the coupling sections, a technique which is useful with narrow-band filters. This is discussed in greater detail in Chapter 4.

The performance which can be obtained with wide-band filters is worthy of consideration. Unfortunately, we have not built any experimental models, although evanescent mode filters with bandwidths of 50% are commercially available [8]. Figure 3.12 shows the analysis of a filter of bandwidth approximately 2.3 GHz designed using the current theory and intended for operation at 4 GHz. This filter was designed for a ripple factor of 0.01 dB, but the detailed performance shown in the inset of Fig. 3.12 indicates that this ripple is exceeded over a large part of the band. Nevertheless, the calculated voltage standing wave ratio (VSWR) at worst only marginally exceeds 1.5:1 which is a commonly accepted maximum in filter specifications. These curves represent the performance of a design before any bandwidth equalization routines are employed to improve it. Filters designed for a 1 GHz bandwidth at this frequency realize computed VSWRs not exceeding 1.25:1. Thus the theory can be used to design filters with reasonably wide bandwidths. However, this is an area which would be worth further investigation. A paper by Hupert [9] on wideband filters in cutoff guides is of interest in this connection.

References

1 Craven, G. F. and Mok, C. K. The design of evanescent mode waveguide filters for a prescribed insertion loss characteristic, *IEEE Trans. Microwave Theory Tech.*, **MTT-19** (3), 295, March 1971.

2 Shea, T. E. *Transmission Networks and Wave Filters*, Fig. 160B, Van Nostrand, Princeton, NJ, October 1929.

3 Cohn, S. B. Direct-coupled resonator filters, *Proc. IRE*, **45**, 187, February 1957.

4 Snyder, R. V. New application of evanescent mode waveguide to filter design, *IEEE Trans. Microwave Theory Tech.*, **MTT-25** (12), 1013, December 1977.

5 Craven, G. Waveguide bandpass filters using evanescent modes, *Electron. Lett.*, **2**, 251, 1966.

6 Nicholson, B. F. and Powell, I. L. Equivalence between evanescent-mode and combline filters, *Electron. Lett.*, **3**, 495, 1967.

7 Craven, G. Relationship between direct-coupled waveguide filters and evanescent-mode filters, *Electron. Lett.*, **4** (3), 44, 1968.

8 *Catalog*, pp. 34–36, K. & L. Microwave Inc., 408 Coles Circle, Salisbury, MD 21801, USA, 1983.

9 Hupert, J. J. Evanescent mode coupling in commensurate microwave filters, *Proc. Natl Electron. Conf.*, **26**, 657, December 1970.

Chapter 4

Advanced Filters

4.1 Introduction

Although the design theory described in Chapter 3 satisfied a wide range of system requirements, increasing applications for this type of filter necessitated its extension in several ways. The purpose of this chapter is to describe some of the more sophisticated filters that resulted. These developments took a number of forms and included techniques for shortening narrow-band filters, methods of multiplexing filters of this type in diplexers, triplexers and quadruplexers (in which the filters are effectively in parallel) and a brief investigation of the more general problem involving the constant-resistance channeling filter. Finally, the extension of the techniques used by Atia and Williams [1] in their well-known dual-mode elliptic-function filters to the corresponding below-cutoff counterpart was investigated. These topics are discussed in this chapter.

4.2 Inductive loading [2]

One of the advantages of the present filter type lies in its compactness, which arises from its efficient volume-to-Q-factor ratio. In narrow-band filters this advantage is eroded by the length of the filter, which by comparison with the other dimensions can be excessive. A solution to this problem can be sought by dividing the length between the resonators into two regions: the resonator region and the coupling region. Obviously, this is only qualitatively true because in a distributed-constant medium the transition from one state to the other will be gradual, but it has a general validity because of the different properties that are necessary in the two regions. In the resonator region the storage fields are quite intense and small amounts of loss have a large effect on the overall loss. In the coupling region the storage fields are not intense, and a lower Q factor is permissible here without incurring significant loss. It is, of course, the undue extension of the latter region which is responsible for the increased filter length. It therefore follows that in a primarily inductive medium we can replace part of the high-Q distributed-constant section (the waveguide) with an equivalent

51

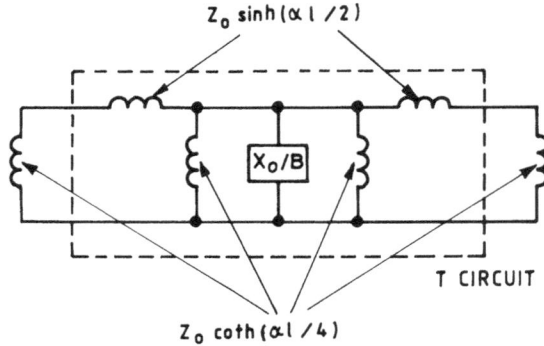

Fig. 4.1   Equivalent circuit of a cutoff waveguide with shunt susceptance.

lumped-circuit element: an inductive obstacle. The equivalent circuit which then represents the two coupled resonators is that of Fig. 4.1.

4.2.1   Design theory

The basic approach to the theory involves designing the filter in its unmodified form, which yields the coupling length l. We then have to determine the modified length l' that is necessary to obtain the same coupling between resonators in the presence of the inductive loading Z_0/B where B is the normalized susceptance. The first step in this procedure is to obtain the π-section that is equivalent to the T-section between the dotted lines in Fig. 4.1. Applying the transformation relating these two networks we have

$$X_a = Z_0 \left[2 \sinh\left(\frac{\alpha l'}{2}\right) + \sinh^2(\alpha l')\left\{B + 2\tanh\left(\frac{\alpha l'}{4}\right)\right\}\right] \tag{4.1}$$

$$= Z_0 \left\{\sinh(\alpha l') + B \sinh^2\left(\frac{\alpha l'}{2}\right)\right\} \tag{4.2}$$

$$X_b = Z_0 \left\{\sinh\left(\frac{\alpha l'}{2}\right) + \frac{2}{B + 2\tanh(\alpha l'/4)}\right\} \tag{4.3}$$

The first necessary condition which must be satisfied is clearly

$$X_a = Z_0 \sinh(\alpha l) \tag{4.4}$$

where the term on the right represents the coupling inductance in an unmodified filter. This leads to a value for B of

$$B = \frac{2\{\sinh(\alpha l) - \sinh(\alpha l')\}}{\cosh(\alpha l') - 1} \tag{4.5}$$

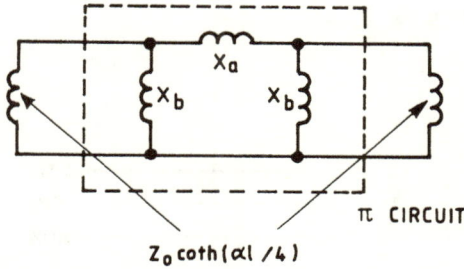

$Z_0 \coth(\alpha l /4)$

Fig. 4.2 The T-circuit of Fig. 4.1 transformed to a π-circuit.

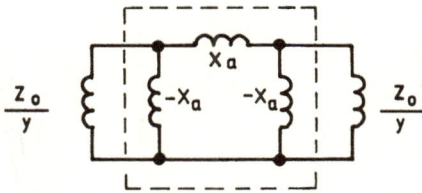

Fig. 4.3 Equivalent circuit with a J inverter.

resulting in the equivalent π-section within the dotted lines in Fig. 4.2. We now have to rearrange the element values in Fig. 4.2 so that the sum of the shunt elements remains unchanged but the element inside the dotted rectangle of Fig. 4.3 is equal to $-X_a$. Thus we have the relation

$$\frac{1}{Z_0 \coth(\alpha l'/4)} + \frac{1}{X_b} = \frac{Y}{Z_0} - \frac{1}{X_a} \qquad (4.6)$$

or

$$Y = \coth(\alpha l') \frac{1 + (B/2)\tanh(\alpha l')}{1 + (B/2)\tanh(\alpha l'/2)} \qquad (4.7)$$

Obviously, if l' is reasonably long $\tanh(\alpha l'/2) \rightarrow \tanh(\alpha l')$ and the resonator inductance in Fig. 4.3 remains virtually unchanged. For substantial reductions in length (50%) some change occurs, principally as a result of the change in resonator inductance and the correction factor. This is illustrated in the curve of Fig. 4.4 which compares the computed bandwidths for various degrees of reduction. In the original paper [2] a more refined design theory is developed which takes into account the changes in Y which occur when l' is substantially reduced. However, this theory involves an iterative solution and still does not produce an exact result. Thus the option exists of employing the approximate

TRANSMISSION LOSS
dB

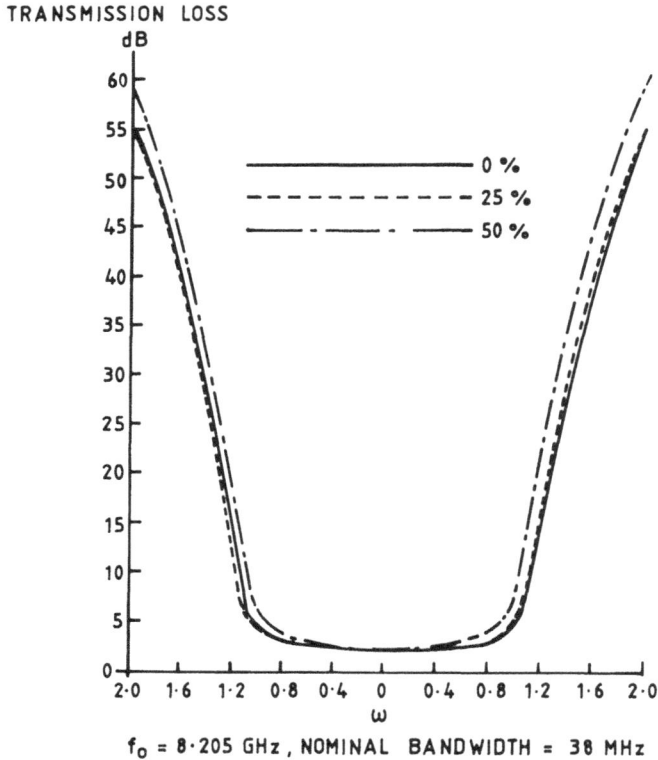

f_0 = 8·205 GHz , NOMINAL BANDWIDTH = 38 MHz

Fig. 4.4   Computed transmission response of a seven-section 0.1 dB Chebyshev filter for various percentage reductions in length.

analysis described above and then subjecting the result to an analysis program and making the required corrections in that way.

4.2.2   Performance

The performance illustrated in Figs 4.5–4.7 illustrates the effect of loading on loss: 25% reduction leads to no measurable increase; the loss is actually lower with 40% reduction, but when allowance is made for the slightly wider bandwidth a small increase in loss occurs. A practical point worth noting is that, where an exact bandwidth must be realized, the expedient employed with direct-coupled above-cutoff filters of placing a small capacitive screw in shunt with the inductive post is quite acceptable. In the latter case the filter is initially designed to have a somewhat narrower bandwidth than the desired value and the difference is taken up in the shunt capacitive screws. The filter was designed for an equal-ripple bandwidth (0.005 dB) of 33.4 MHz. The overall dimensions of the shortest filter constructed in a WR62 square guide were 15.87 cm × 1.78

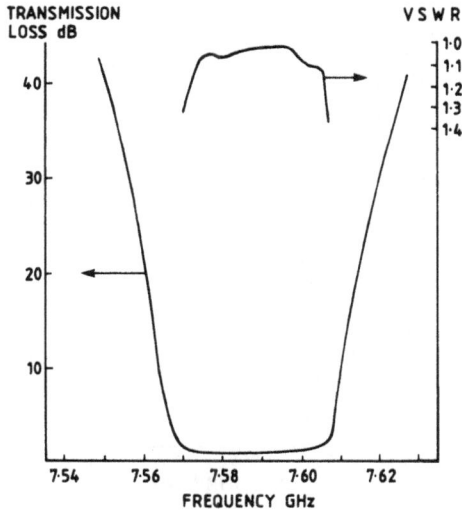

Fig. 4.5   Measured response; midband loss, 1.2 dB.

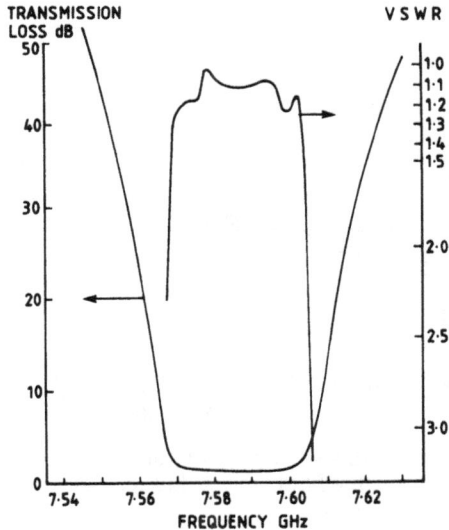

Fig. 4.6   Measured response of a filter with length reduced by 25%; midband loss, 1.2 dB.

cm × 1.78 cm (6.25 in × 0.702 in × 0.702 in). Except that the waveguide size was different the construction was similar to that shown in Fig. 3.10. Quartz-dielectric-rod tuning elements with a slightly smaller diameter were employed. The filter did not include a harmonic rejection filter. In the shortest model the

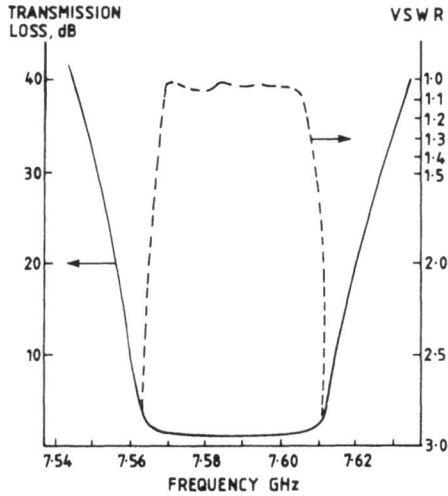

Fig. 4.7   Measured response of a filter with length reduced by 40%; midband loss, 1.11 dB.

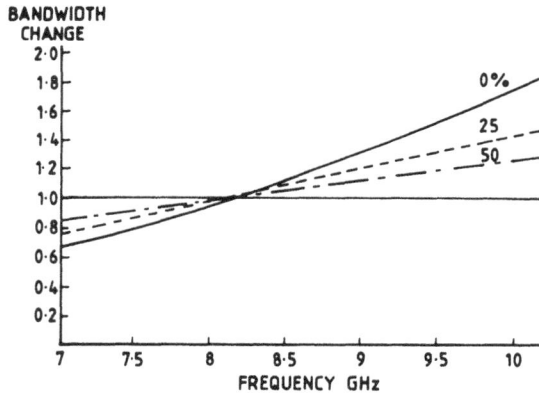

Fig. 4.8   Change in bandwidth with tuning for various degrees of inductive loading for the filters shown in Fig. 4.4.

loading susceptances were obtained with post doublets; thin centered posts were satisfactory in the 25% reduced model. One of the effects of inductive loading is to reduce the variation in bandwidth of the same filter when it is tuned to a different center frequency (Fig. 4.8). It will be seen that with significant loading the improvement in this respect is considerable. Compared with good quality production waveguide filters of the same specifications the loss was 0.2 dB worse, the loss of the latter filters being 1 dB.

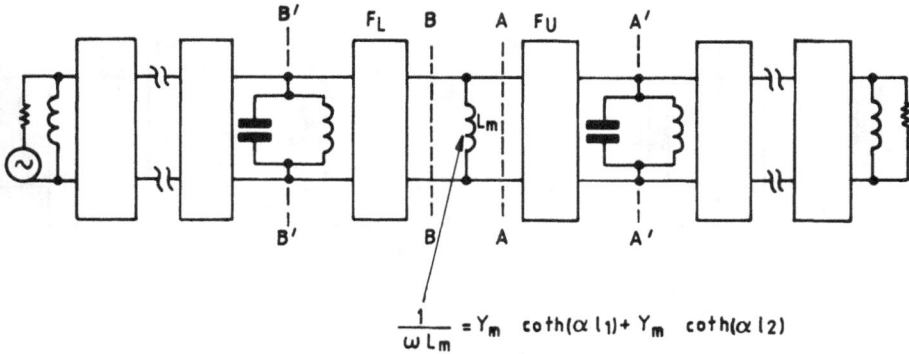

$$\frac{1}{\omega L_m} = Y_m \coth(\alpha\, l_1) + Y_m \coth(\alpha\, l_2)$$

Fig. 4.9 Filters with a common input port.

4.3 Diplexers and multiplexers

Multiplexers involve the connection of two or more filters in parallel (or series) at a common input port, thereby placing the input admittance of each filter in shunt with the other. Mok [3] has described the design theory required for the special case when the filters have virtually contiguous passbands, but the more general problem of filters in which a significant guard band exists (which may be quite small) is considered here. In its simplest form the problem is represented by the diplexer in a cutoff guide illustrated in Fig. 4.9. The essential requirement is to nullify the admittance of the complementary filter, which in its stopband is virtually a pure susceptance, in the passband of the other filter. In addition, the real part of the filter passband impedance must be matched to the source, and in the present problem these impedances are not necessarily the same, i.e. the filters may not be complementary.

4.3.1 Design theory

We designate the upper-frequency and lower-frequency filters as F_U and F_L respectively. The corresponding admittances Y_U (at ω_U) and Y_L (at ω_L) must match the transformed source admittance Y_0 at the band center angular frequencies. Thus in the example where the filters are complementary,

$$Y_U = Y_0 = Y_L \qquad (4.8)$$

the imaginary part of the admittance at the band center must be zero and the conductance terms can be equated:

$$G_U = G_0 = G_L \qquad (4.9)$$

Clearly, the input circuit must resonate at ω_U and ω_L, which of course is possible because at the first of these angular frequencies F_L is an inductive

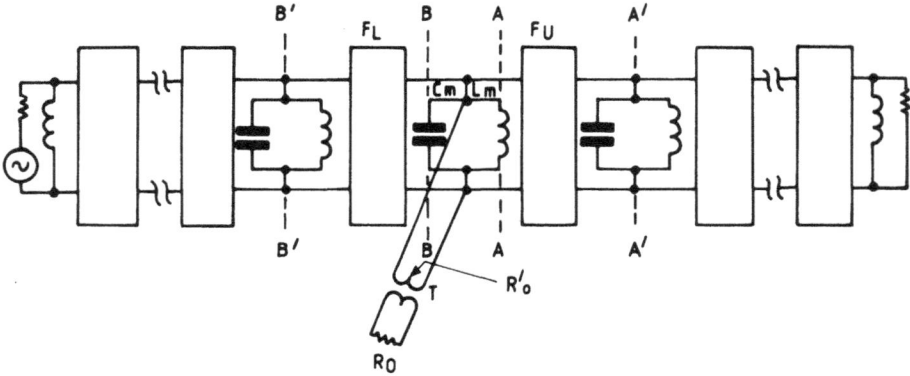

Fig. 4.10   Theoretical diplexer circuit.

susceptance and at the second frequency it presents a capacitive susceptance. The second condition which must be satisfied is that the input conductance presented by the filter (transferred from its load) must match the source conductance (or in the example of an even-order filter must mismatch it by a prescribed amount). Both these conditions have to be satisfied by the coupling length — the $\sinh^2(\alpha l)$ term. Obviously a second variable is essential in order to make this possible; this is supplied by the additional capacitance placed in shunt with the combined remnant inductances of the two filters, as shown in Fig. 4.10. The way in which it is achieved is described in the following sections.

First, we summarize the design problem as it appears so far. The resonance condition for each of the band center angular frequencies ω_U and ω_L is determined by the central resonator frequency ω_m (L_m and C_m) plus the transformed out-of-band susceptance of the filter which is defined as commencing at the last resonator of the filter. Similarly, these transformers, which are the connecting lengths to the first resonator, will control the transformed load resistance represented at the terminals of this filter in its passband. Both these conditions have to be satisfied; the variables available for the purpose are l_1, l_2 and ω_m, of which the variable part is C_m. This leads to the equations

$$\frac{1}{L_m(C_m + C_\delta)} = \omega_L^2 \tag{4.10}$$

$$\frac{L_m + L_\delta}{L_m C_m L_\delta} = \omega_U^2 \tag{4.11}$$

which express the two resonant frequencies in terms of appropriate parameters. The magnitude of this capacitive susceptance is given by

$$\omega_L C_\delta = \frac{1}{Z_L^2 \sinh^2(\alpha_L l_1) B_1} \tag{4.12}$$

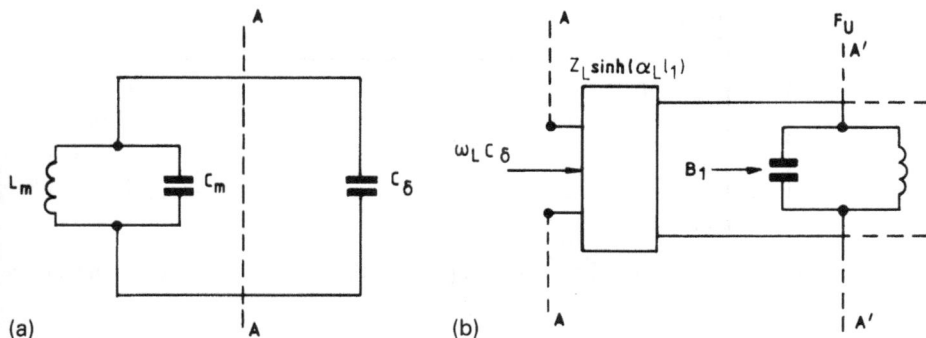

Fig. 4.11 (a) Resonant circuit at ω_L; (b) input susceptance at ω_L of F_U at plane AA.

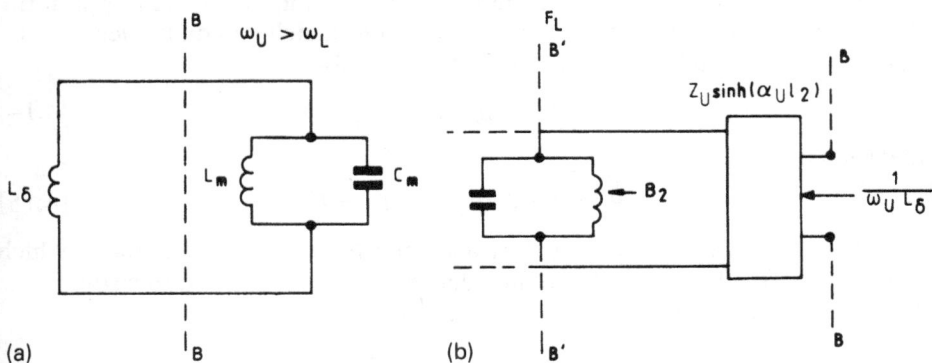

Fig. 4.12 (a) Resonant circuit at ω_U; (b) input susceptance at ω_U of F_L at plane BB.

where Z_L and α_L are respectively the wave impedance and the attenuation constant of the waveguide at ω_L. B_1 is the inductive susceptance of F_U at ω_L in the plane A'A (Figs 4.11(a) and 4.11(b)) inverted to form the equivalent capacitive susceptance $\omega_L C_\delta$.

Similarly, the magnitude of the inductive susceptance at ω_U is

$$\frac{1}{\omega_U L_\delta} = \frac{1}{Z_U^2 \sinh^2(\alpha_U l_2)\, B_2} \tag{4.13}$$

where the subscript U denotes the value of Z at ω_U and B_2 is the corresponding capacitive susceptance of the filter F_L at ω_U determined at plate B'B' in Figs 4.12(a) and 4.12(b) and similarly inverted at the central resonator.

The reason for choosing the above reference planes lies in the ease with which the input impedance of the filter can be defined at that plane. The filter can be designed using the methods described in Chapter 3 as one starting with a

$Z_L \sinh(\alpha_L l_2)$ $Z_U \sinh(\alpha_U l_1)$

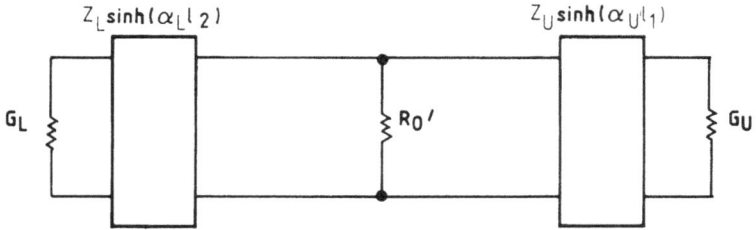

G_L R_0' G_U

Fig. 4.13   Condition for matching the real parts of the filter impedances to the source R_0'.

shunt resonator, and the input impedance can be calculated at any desired frequency.

Corresponding equations relate the resistive components that exist in the passband. For example, if the filter is designed to match an admittance G_U, the impedance at the end of the line l_1 will be (Fig. 4.13)

$$R_U = Z_U^2 \sinh^2(\alpha_U l_1) \; G_U = R_0' \qquad (4.14)$$

and similarly

$$R_L = Z_L^2 \sinh^2(\alpha_L l_2) \; G_L = R_0' \qquad (4.15)$$

Equations (4.14) and (4.15) provide the fundamental constraints which must be observed if the midband impedances are to be matched to a source R_0. Clearly, the lengths will be related in the ratio

$$\frac{\sinh^2(\alpha_U l_1)}{\sinh^2(\alpha_L l_2)} = \frac{Z_L^2 G_L}{Z_U^2 G_U} \qquad (4.16)$$

Equations (4.12) and (4.13) indicate that there is considerable freedom in the choice of the magnitudes of l_1 and l_2 that can be employed, but eqn (4.16) constrains the ratio that these l values may have to each other. At first sight this may appear to be a severe restriction, but this is not so because we have the variable $L_m C_m$ product. In principle, ω_m can have any value between zero and infinity, i.e. when $C_m \to \infty$ and when $C_m \to 0$. In practice the range of interest is

$$\omega_L < \omega_m < \omega_U$$

If we consider the limiting condition within this range when

$$\omega_m \to \omega_L$$

then from eqn (4.12) only a small value of C_δ is required. This value, in turn, can only be met by a larger value of $Z_L \sinh(\alpha_L l_1)$, which means that the necessary matching conditions (4.14) and (4.15) can be satisfied with very different values of G_U and G_L. This corresponds to the condition in which filters with very different characteristics — a broad-band and a narrow-band filter — can be matched to the same source impedance. Thus the theory provides a technique which enables a wide variety of filters to be used in a diplexer. The freedom is the result of the variable capacitance placed across L_m and the coupling loop. The latter removes the constraint that the filters must match a specific source impedance at their mutual junction, although it is essential that they match an equal, or approximately equal, impedance.

The second limiting condition needs to be considered, i.e.

$$\omega_m \rightarrow \omega_U$$

This is merely a reversal of the previous situation, and in this case the wide-band filter is the low-frequency filter. Finally, the condition in which ω_m is located at the geometric mean of ω_L and ω_U is of interest because it represents the situation in which the two filters have identical characteristics except for their center frequencies. This illustrates the importance of ω_m in the problem and we therefore rearrange (4.10) and (4.11) in these terms. Since $\omega_m^2 = 1/L_m C_m$ we have

$$\frac{1}{\omega_m^2} = \frac{1}{\omega_L^2} - L_m C_\delta \tag{4.17}$$

$$\frac{1}{\omega_m^2} = \frac{L_m + L_\delta}{\omega_U^2 L_\delta} \tag{4.18}$$

On multiplying (4.17) by $1/C_\delta$ and (4.18) by $L_\delta \omega_U^2$ and adding the two equations we get

$$\omega_m^2 = \frac{\omega_L^2(1 + \omega_U^2/\omega_\delta^2)}{1 + \omega_L^2/\omega_\delta^2} \tag{4.19}$$

where the product $L_\delta C_\delta$ is conveniently represented by $1/\omega_\delta^2$. However, combining (4.12) and (4.13) gives

$$\omega_\delta^2 = \frac{\omega_U \omega_L B_1}{B_2} \left\{ \frac{Z_L \sinh(\alpha_L l_1)}{Z_U \sinh(\alpha_U l_2)} \right\}^2 \tag{4.20}$$

This is similar to expression (4.16) relating G_L and G_U, and since $Z = j\omega\mu/\alpha$ the expansion for $Z \sinh(\alpha l)$ is given by

$$Z \sinh(\alpha l) = \frac{j\omega\mu}{\alpha} \left\{ \alpha l + \frac{(\alpha l)^3}{3!} + \ldots \right\}$$

Clearly, for small values of αl, α cancels in the expansion and under these conditions

$$\frac{Z_L \sinh(\alpha_L l_1)}{Z_U \sinh(\alpha_U l_2)} \approx \frac{Z_U \sinh(\alpha_U l_1)}{Z_L \sinh(\alpha_L l_2)} = \left(\frac{G_L}{G_U} \right)^{1/2} \tag{4.21}$$

This approximate relationship can be used as the basis of a solution which, in general, will converge quite rapidly. Substituting (4.21) in (4.20) we have

$$\omega_\delta{}^2 \approx \frac{\omega_U \omega_L B_1 G_L}{B_2 G_U} \tag{4.22}$$

Since the terms in (4.22) represent the out-of-band admittance of each filter at the design frequency of the other, and also the conductance values required to match the filter at the defined reference planes, the equation is an explicit (approximate) solution for ω_δ. Consequently, if we substitute (4.22) in (4.19) we can determine the value of ω_m. It is then only necessary to substitute the value of ω_m in (4.10) and (4.11) to obtain C_δ and L_δ. The magnitudes of the electrical lengths follow from these values of $\sinh(\alpha l_1)$ and $\sinh(\alpha l_2)$ (eqns (4.12) and (4.13)).

For many practical purposes the foregoing can be embodied in two approximate equations which yield the lengths directly:

$$\sinh(\alpha_L l_1) \approx \frac{1}{Z_L} \left\{ \frac{\omega_m L_m}{B_1(\omega_m/\omega_L - \omega_L/\omega_m)} \right\}^{1/2} \tag{4.23}$$

$$\sinh(\alpha_U l_2) \approx \frac{1}{Z_U} \left\{ \frac{\omega_m L_m}{B_2(\omega_U/\omega_m - \omega_m/\omega_U)} \right\}^{1/2} \tag{4.24}$$

4.3.2   Design summary

Although the design theory looks complicated the design procedure can be summarized briefly.

(1) Design filters F_U and F_L as filters which commence as a parallel resonator. Then, with the filter terminated in its far end in the correct load and using a suitable program, compute B_1, G_U, B_2 and G_L.

(2) Compute ω_δ using eqn (4.22).

(3) Compute ω_m using eqn (4.19).

(4) Finally, determine $\sinh(\alpha l_1)$ (and l_1) and the corresponding values for l_2. These are found from eqns (4.23) and (4.24). A design program is necessary because of the essentially iterative nature of the solution: the susceptance of L_m depends on the connecting length, which is in turn influenced by the magnitude of the central shunt resonator.

4.3.3   Performance

In order to test the theory a diplexer was designed to meet the provisional specifications required in a maritime satellite. Analysis of the proposed structure confirmed the accuracy of the design and it was therefore built. The construction of the diplexer is shown in Fig. 4.14. Prior to construction the two

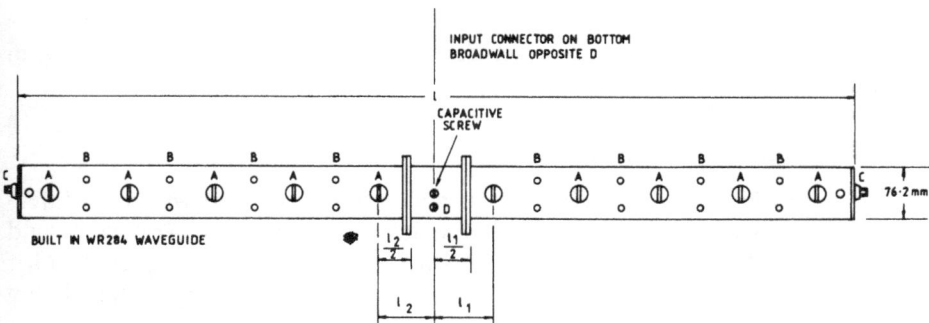

Fig. 4.14 Diplexer construction: A, capacitive rod; B, inductive posts; C, connector; D, coupling adjustment; $l_1 = 112$ mm; $l_2 = 102$ mm; $l = 1296$ mm.

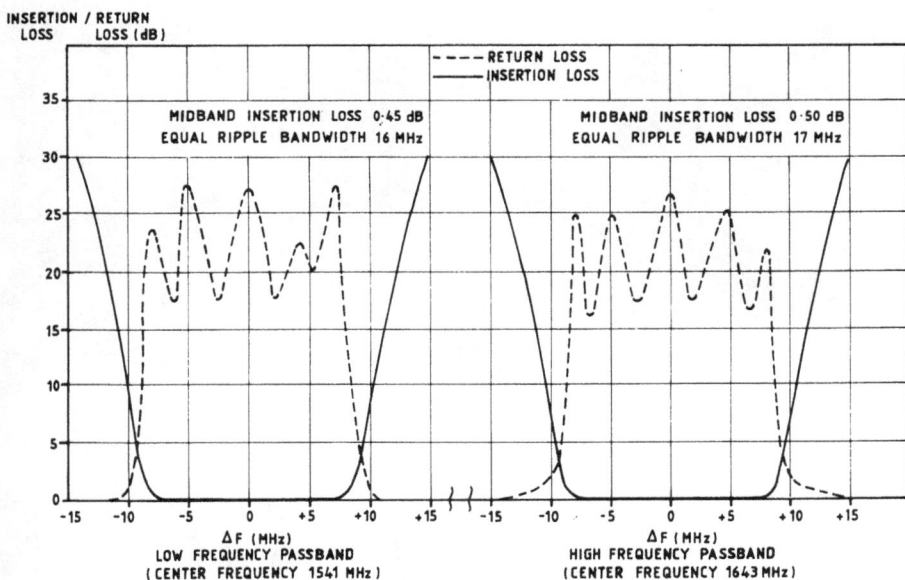

Fig. 4.15 Transmission characteristics of a diplexer in the passband region.

filters were tested independently as separate units in order to establish criteria for judging the performance of the ultimate diplexer. The performance shown in Fig. 4.15 indicates virtually identical results; in particular, the 0.1 dB ripple factor is closely maintained. The loss, which was in the region of 0.5 dB for both filters, is typical of uncleaned copper cavities of this type constructed in a rectangular guide.

The foregoing diplexer was built in anticipation of the contract for the Marecs satellite, partly to ensure that an adequate theory was available. The ultimate model involved a number of trade-offs between loss, size and power-handling capacity and has been described in the literature [4]. It is discussed in Chapter 8.

4.3.4 Additional applications

The flexibility of the theory, with its ability to match filters with very different characteristics into a common load, invites wider uses. Triplexers and quadruplexers have both been built; a photograph of and performance data for the latter are shown in Figs 4.16 and 4.17. The star-type configuration shown

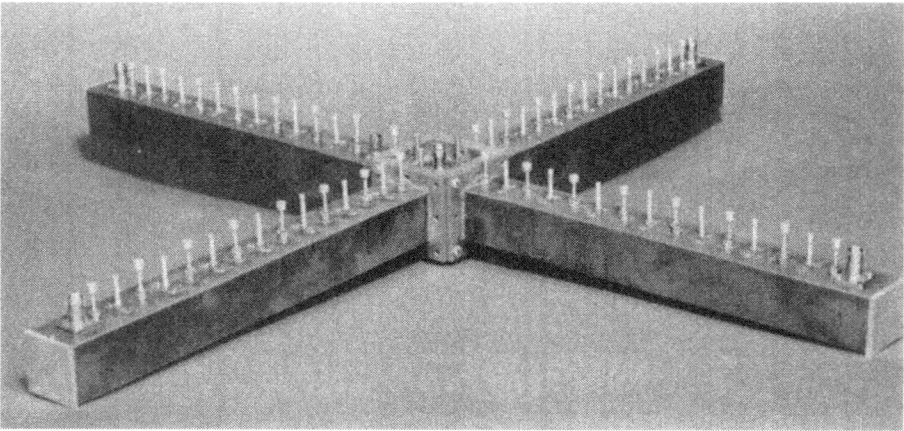

Fig. 4.16   A quadruplexer.

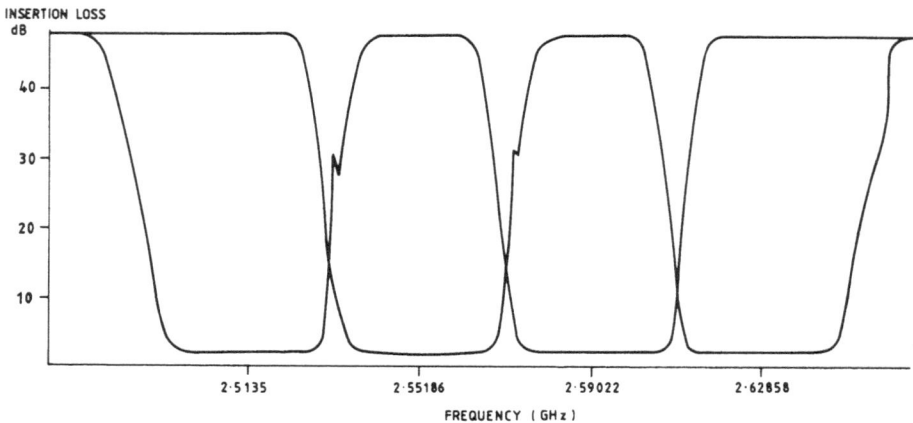

Fig. 4.17   Performance of a quadruplexer.

was convenient for the systems involved but is not absolutely essential. The design theory can also be applied to other subsystems. These include parametric up-converters and microwave receiver heads. Harmonic multipliers are also attractive, but some additional factors have to be considered here. These matters are also discussed further in Chapter 8.

4.4 Directional filters

4.4.1 Introduction

The filters described in the previous section can be broadly termed channeling filters as their function is to channel carriers from (or to) a given antenna to (or from) the various amplifiers with which they are associated. The individual carriers may represent a multichannel telephone system, television or data, all of which have a common destination. Techniques for multiplexing in this way have grown popular in satellite systems because they are lightweight, low loss and economical. Their success in such systems derived from these characteristics, but because of the said interaction between the various filters it is essential that the multiplexing unit be designed and tested as an integral unit. Thus it is not practicable to add units later without redesigning the complete subsystem. This disadvantage is of small account in a satellite once it is launched, but in ground station systems, especially microwave telephone links, it can be a serious disadvantage. For example, at some stage after a complete link has been installed it may be necessary to add a spur link required by an increase in local traffic. If the multiplexing techniques described in the previous section were used, this would necessitate much expensive redesign and installation. In these circumstances the constant-resistance, or directional, filter represents a better solution because of its greater flexibility.

4.4.2 Types of directional filters

One of the most popular directional filter arrangements is illustrated in Fig. 4.18. It consists simply of bandpass filters connected to the second part of a three-port circulator. The mode of operation is largely self-evident: signals that are not in the passband of the first filter in the chain are reflected back to the circulator and appear at the port of the next filter, each filter dropping off one channel at a time. The scheme has the advantage of flexibility and is also reasonably cheap. The disadvantage is, of course, the additional circulator loss, which in a long chain can easily exceed 1 dB for the last channel.

The basic directional system is illustrated in Fig. 4.19. It consists of two hybrid-Ts and two filters. Bandpass filters are shown in the illustration, although in the original design [5] band-rejection filters were employed. The operation is as follows. Power fed into the hybrids divides equally between the two arms. At

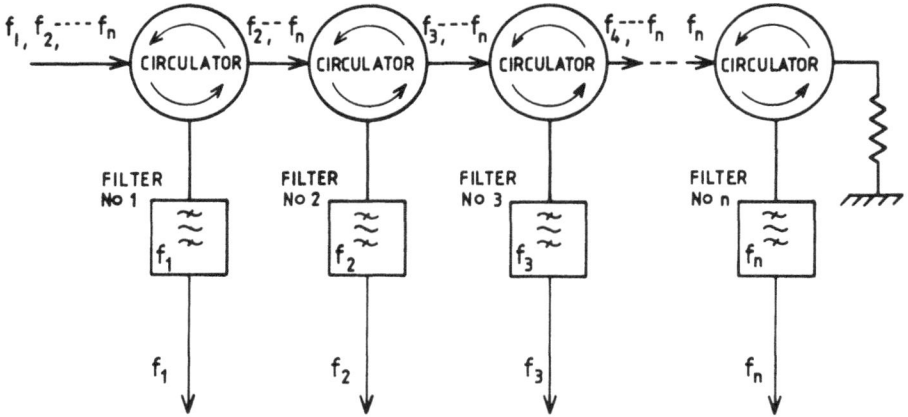

Fig. 4.18 Conventional channel-dropping circuit using circulators.

frequencies in the passband of the filters power travels through the identical filters and recombines in the corresponding arm of the hybrid. Outside the passband power is reflected back to the input hybrid and, because of the total 180° shift in one arm, the output appears at the fourth arm of the hybrid. If 3 dB directional-coupler hybrids are used a 90° phase shift is intrinsic in the system. The change to band-rejection filters interchanges the ports at which the dropped channel and the channels passed on to the remaining filters appear. The mass of the two hybrid junctions and the two filters is considerable. This is partly mitigated in the dual-mode directional filter, but the weight of the circular guide and the two output guides is still substantial. The dual-mode evanescent mode directional filter in which the input ports consist of coaxial input connectors greatly improves the situation.

4.4.3 Dual-mode evanescent mode directional filter

Consider a conventional hybrid junction as shown in Fig. 4.20. The equivalent junction in a cutoff guide can be illustrated using dual-mode resonators optimally coupled to the structure shown in Figs 4.21(a)–4.21(c). First, regarding the structure in its constituent parts, we have in Fig. 4.21(a) the E field incident on the wire shown. Clearly, this field will induce a current in the wire which is equivalent to a generator located at the center as shown and current will flow into loads located at output ports 1 and 2. Now consider a corresponding resonator in the orthogonal mode optimally coupled into the vertical wire (Fig. 4.21(b)) and joining the horizontal wire at its midpoint. Provided that ports 1 and 2 are matched the currents in the two halves of the horizontal wire will be

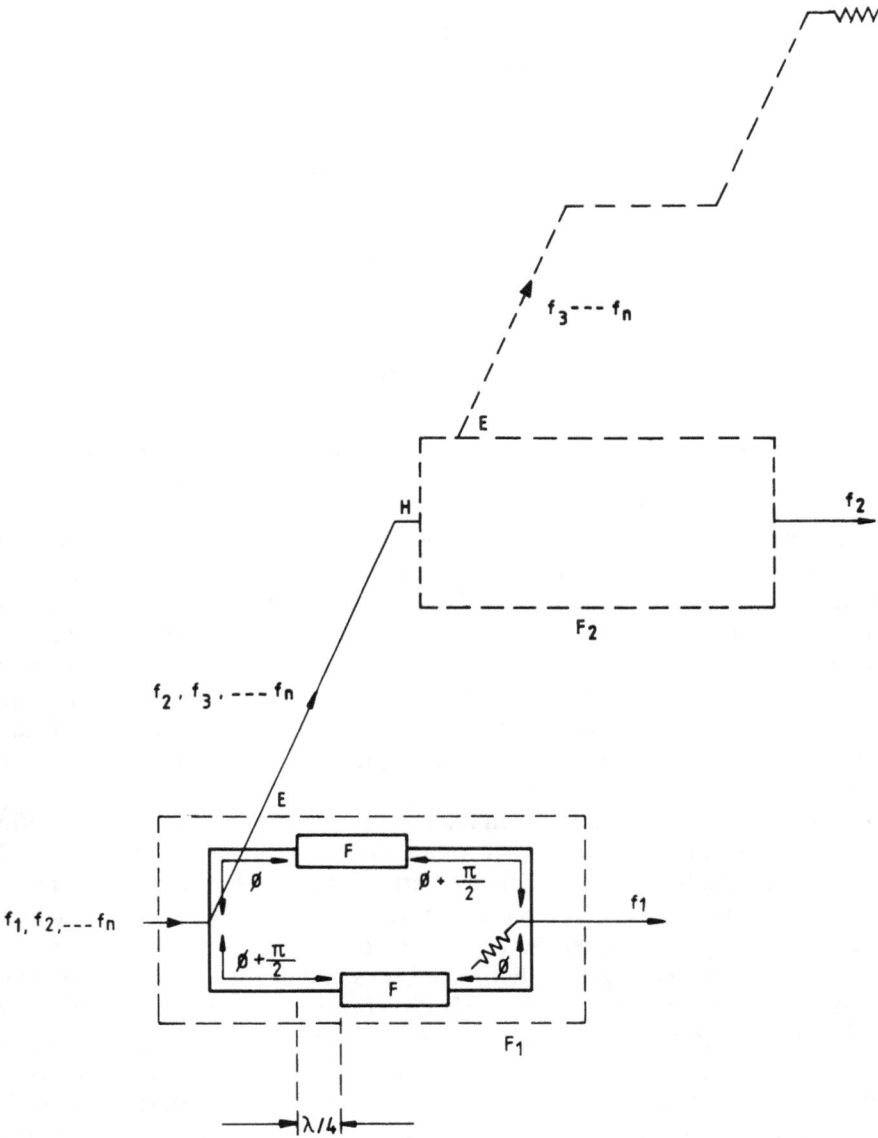

Fig. 4.19 Cascaded directional filters.

equal and opposite. They will therefore cancel in any coupling to the TE_{10} mode. Thus we have the basic properties required in a hybrid-T. It is only necessary to ensure that the output ports are correctly matched and that we have a hybrid-T at each end of the dual-mode filter (Fig. 4.21(d)).

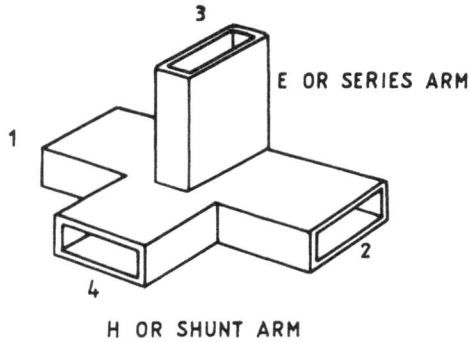

Fig. 4.20   Conventional hybrid-T.

The input impedance that such a structure presents at its terminals is important for matching to the outside cable. The short circuit at the center makes it fundamentally a highpass structure and therefore it is important that the structure presents a real image impedance at its input terminals. With the dimensions involved the network had a high but real image impedance which was due in part to the distance to the back plate, which forms a kind of "ground plane". This impedance can be reduced by a shunt capacitance located at the center of the network. The circulating current in the short-circuit branch, which is caused by this partial resonance, is also beneficial in increasing the coupling to the resonators. The open-circuit line section shown in Fig. 4.21(e), and located at an angle which prevents significant coupling to the fields in the guide, serves this purpose.

Such a hybrid-T can be used in the units required in Fig. 4.19, but it suffers from the disadvantage that the 90° phase shift shown in this figure must be provided physically by an impedance inverter for example. This necessitates an additional resonator in each polarization together with its connecting lengths. Filters of this type have been built, but the additional resonator is a disadvantage. A scheme which is generally preferable is one using 3 dB hybrid in which the phase shift is intrinsic in the 3 dB directional coupler. It has been found empirically that this can be achieved by moving the hybrid through an angle so that it couples into the cutoff guide on the diagonal (Fig. 4.22). The modes that will be excited are those shown in Fig. 4.22(b) and they have been illustrated in the literature [6]. It is shown that these modes can, in fact, be resolved into two TE_{10} modes (Figs 4.22(c)–4.22(f)). The phase relationships (measured) indicate that the modes in the figure represent a circularly polarized wave at specific moments during its rotation.

4.4.4   Performance

Tests were made initially to confirm that the diagonally excited unit had the necessary directional-coupler phase relationships. This is illustrated in Fig.

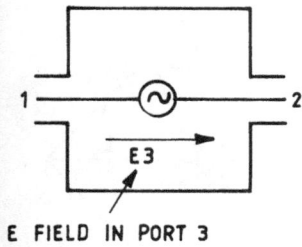

E FIELD IN PORT 3

(a)

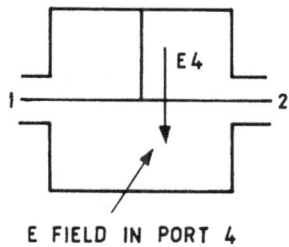

E FIELD IN PORT 4

(b)

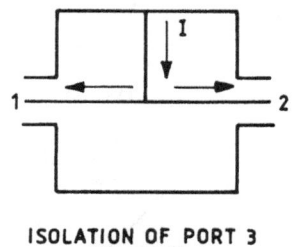

ISOLATION OF PORT 3 AND PORT 4

(c)

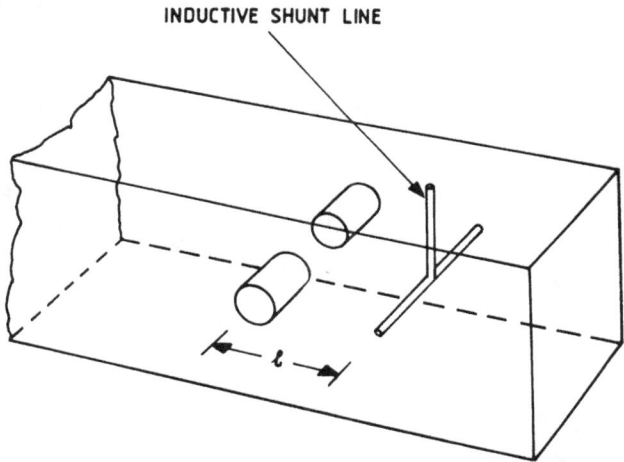

INDUCTIVE SHUNT LINE

UNMATCHED STRUCTURE (SHOWN FOR ONE POLARIZATION)

(d)

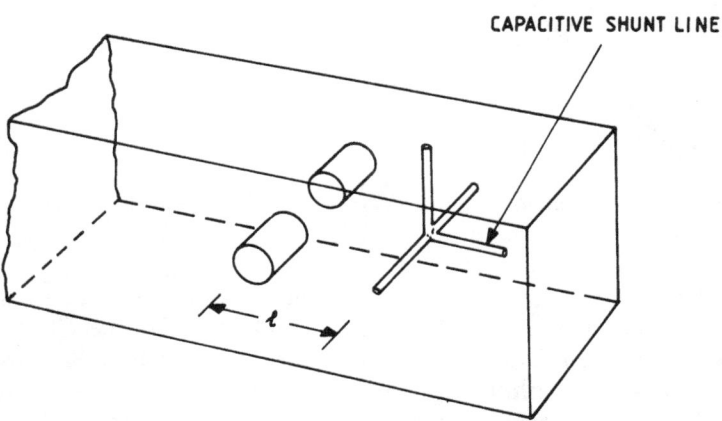

CAPACITIVE SHUNT LINE

STRUCTURE SHOWING CAPACITIVE MATCHING ELEMENT

(e)

Fig. 4.21 Equivalent hybrid-T using a cutoff guide.

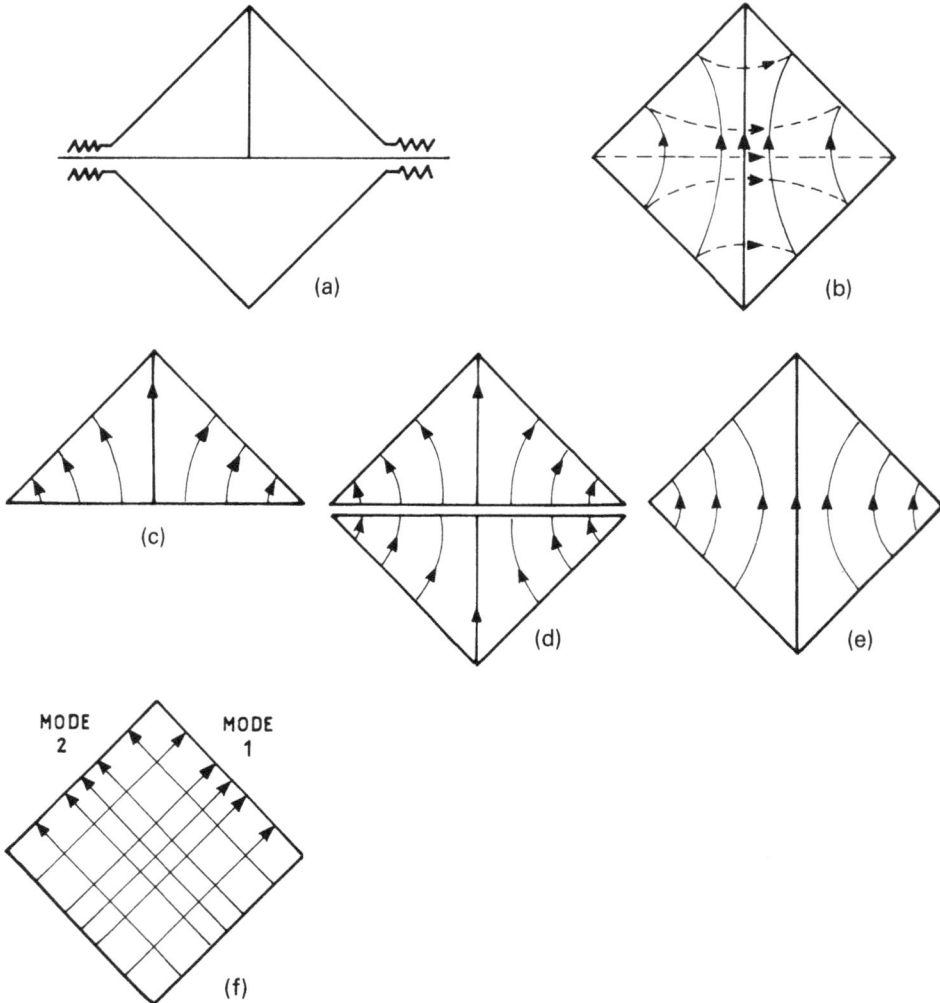

Fig. 4.22   Directional coupler showing resolution into TE_{10} modes.

4.23(a), curve 1, where the resonators are completely detuned. A corresponding hybrid-T would completely reflect the power back to the source. Curve 2 shows the return loss at port 1. The complete performance of the filter is shown in Fig. 4.23(b). Figure 4.24 is a photograph of the two versions of the filter, which was constructed in a square waveguide of dimensions 7.63 cm × 7.63 cm. The weight was less than a quarter of that of a narrow-band propagating guide filter of circularly polarized design. Both filters met the loss specification (0.5 dB); the conventional filter used only two narrow-band cavities in order to achieve the

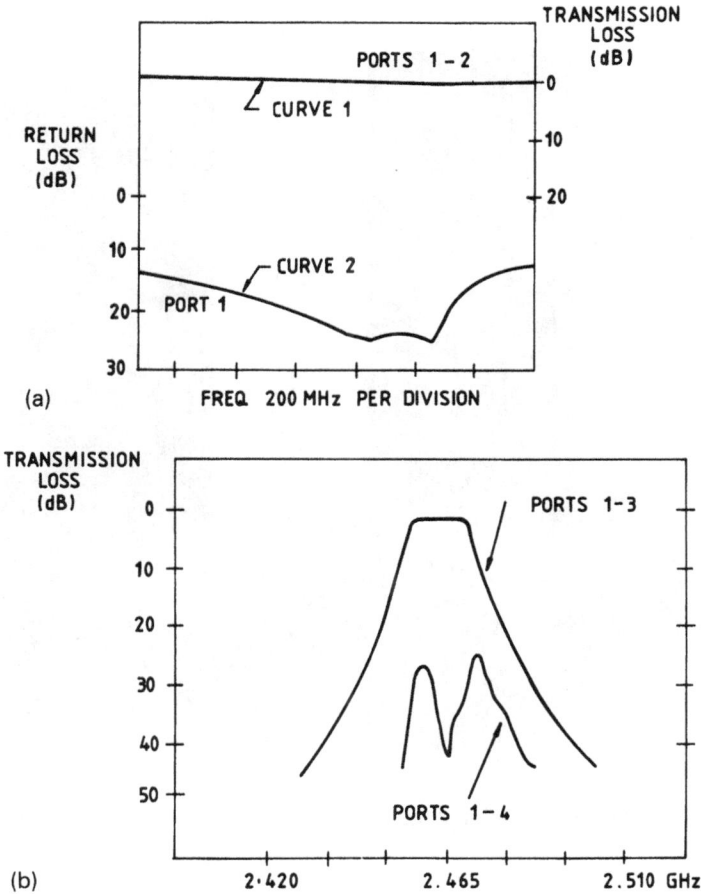

Fig. 4.23 (a) Measurements on a 3 dB coupler with open circuits at the resonators in ports 3 and 4; (b) performance of a three-section directional filter (3 dB coupler version).

same out-of-band rejection. A comparison of the dimensions of the two filters is shown in Fig. 4.25.

4.5 Directional filters employing two filters

The basic configuration for a directional filter employing two filters was first proposed by Cohn and Coale [7] in a single-section resonator version, but the correctness of the theory was later disputed [8] and it was concluded that the

Fig. 4.24   Experimental filters built in a square WR187 waveguide.

filter could not have directional properties. The configuration illustrated in Fig. 4.26 differs slightly from the original configuration of Cohn and Coale: the three-quarter-wave coupling section in the top sector, which provides the required phase reversal, is replaced by a phase reversal in the lower circuit, and the upper and lower halves of the circuit are then of the same electrical length. This phase reversal would not be easily provided in the original circuit but presents no problems in a practical circuit incorporating the present filters. The properties needed if the circuit is to have directional properties can be derived from inspection.

4.5.1   Design conditions

The circuit in Fig. 4.27 represents the directional filter at the band center. Because of the 180° difference in the excitation of the two paths, we can expect their electrical lengths to be equal; the longer path incorporating the two

FILTER TYPE	LENGTH cm	WIDTH cm	MASS kg	LOSS dB
PROPAGATING (2 SECTION NARROW BAND)	58	13 (DIA)	7·1	0·5
EVANESCENT (3 SECTION WIDER BAND)	33	7·63	1·7	0·5

BOTH FILTERS CONSTRUCTED FROM 1·5mm THICK INVAR

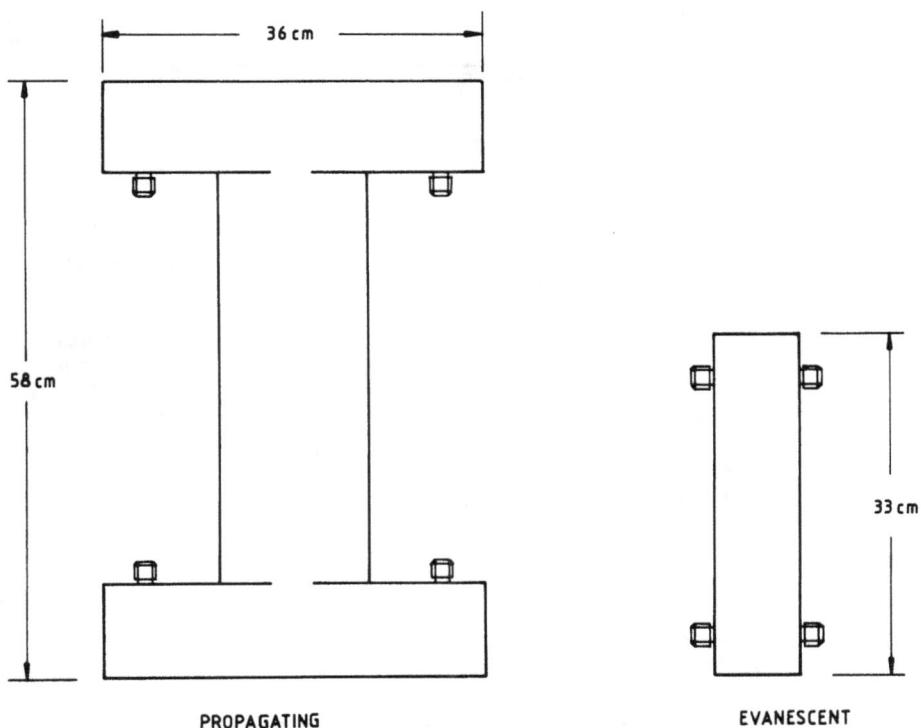

PROPAGATING   EVANESCENT

Fig. 4.25 Comparison of propagating and evanescent mode directional filters.

quarter-wave coupling sections will be exactly cancelled out by the phase inversion at the beginning of the shorter path. Thus the waves will add at port 3 without reflection provided that the internal route is twice the characteristic impedance of the termination as shown. For the same reasons the waves will cancel at port 4, because the electrical length of the two paths are identical but

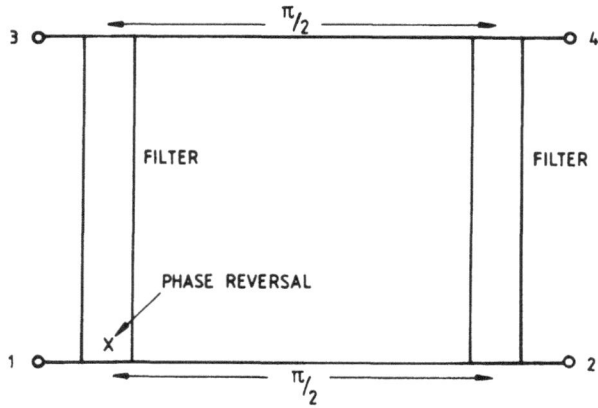

Fig. 4.26   Basic configuration.

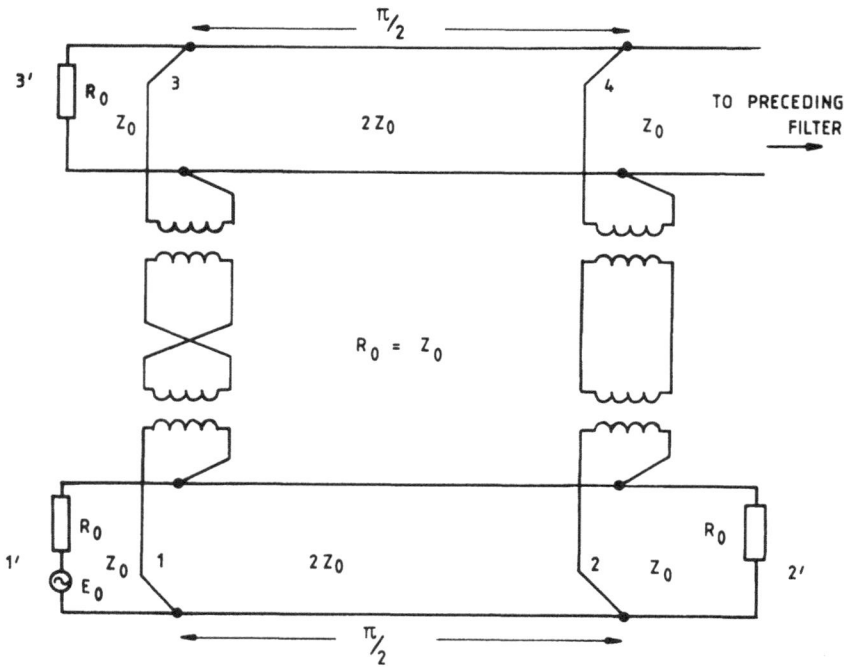

Fig. 4.27   Equivalent circuit in passband.

they are initially excited in antiphase. The condition for cancellation at port 2 is more complex and places constraints on the path lengths of the arms that will contain the two filters. Thus, if we define the lengths of the two paths by ϕ_1 (the shorter path) and ϕ_2 we have

$$\phi_1 = \frac{\pi}{2} \qquad \phi_2 = \frac{\pi}{2} + 2\theta$$

where θ is the phase shift per filter. Consequently, for cancellation to occur at port 2 it is necessary that

$$2\theta = 2n\pi$$

This is the important extra constraint imposed on the filter paths if the complete network is to satisfy the requirements of a directional filter. We now consider two possible configurations: a three-resonator filter in each arm (Fig. 4.28) and a second arrangement comprising two resonators (Fig. 4.29). It will be evident that cancellation of the output will be achieved only with an odd number of resonators. With an even number, the output will be shared between ports 3 and 2.

4.5.2   Performance

The directional filter shown in Fig. 4.28 was built using three resonators per filter plus the output coupling section. The short connecting lengths to the filters were adjusted in practice in order to obtain the correct measured phase path length around the loop. In principle, these could be used to accommodate an even number of resonators in the filters but would have to be much longer. The necessary phase reversal in excitation of the filter units is achieved quite simply by the scheme illustrated in Fig. 4.30. The isolation between ports 1 and 4 is approximately 30 dB or greater; it ripples around 20 dB and decreases to zero outside the passband. The detailed performance is shown in Fig. 4.31.

The advantage of the filter compared with conventional filters in waveguides lies in the elimination of the weighty hybrid-T junctions. Instead the filters are an intrinsic part of a hybrid ring, the remaining part being the interconnecting coaxial lines which can be quite light. Its disadvantage lies in the need for attention to the length of the phase path in the two filters.

4.6   Elliptic function filters

The presentation of a theory on the design of elliptic function filters in a cutoff guide presents some problems, principally because such filters are, self-evidently, variations on the basic theory propounded by Atia and Williams [1].

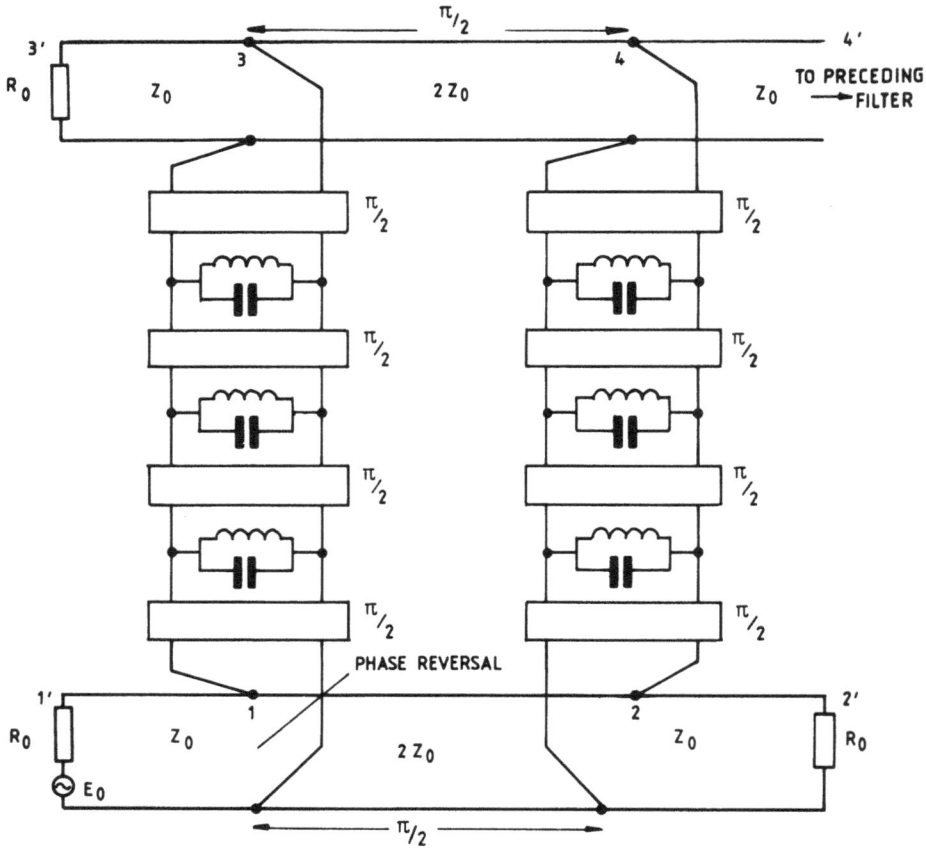

Fig. 4.28   Waves cancel at port 2 (band center).

Thus the approach in a commercial laboratory already dedicated to space research was simply to adopt the procedure of using what was already available. This consisted of modifying the programs already developed for the propagating version so that they could accommodate the changes that are necessary in the cutoff-guide version. Whilst this makes good sense from a practical engineering and commercial viewpoint, it makes the exposition of the subsequent theory quite difficult. The reasons are twofold: first, the complexity of the theory means that the appropriate programs are massive and their inclusion in the present text would be completely beyond the scope of this book. Nevertheless, because of the importance of this filter, and the fact that practical filters have been built using the modified theory [4], it is felt that an effort should be made to describe the required modifications. This necessitates the assumption that the reader already has these programs, which is somewhat unsatisfactory.

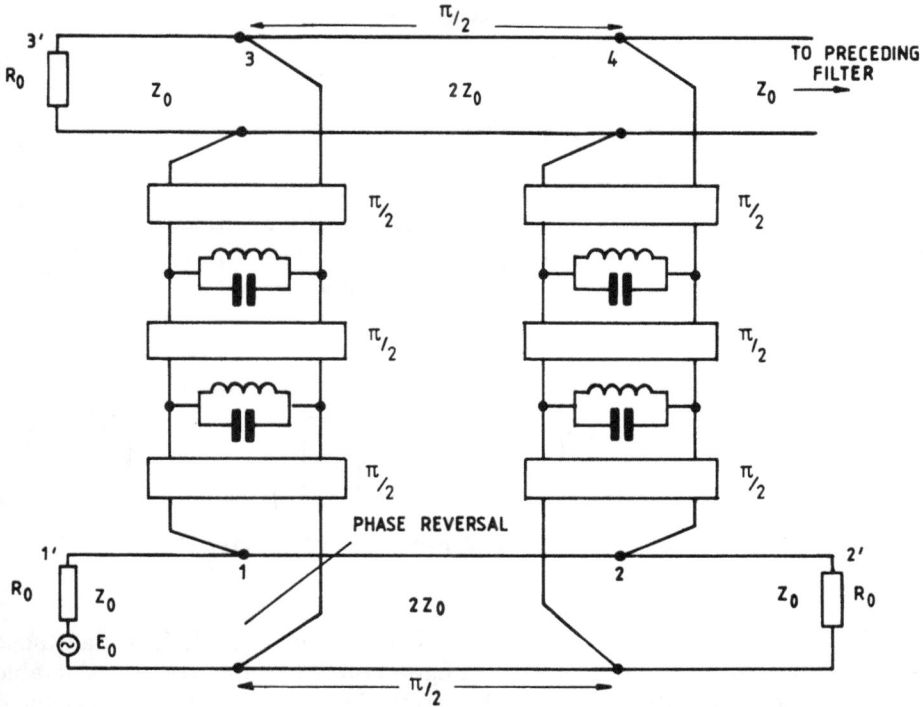

Fig. 4.29 Waves add at port 2 (band center).

However, until a more comprehensive theory is available there seems little alternative. At all events it is hoped that the present exposition will encourage others to pursue the theory further. With these thoughts in mind the following suggestions are offered.

The reason for the complexity of the original theory lies in the generality of the approach [1]: Atia and Williams make few assumptions concerning even the topology of the filter. They initially assume extra couplings between all resonators and then determine, by solving the impedance matrix representing the network, the magnitude of the couplings (including those of zero value) which will yield the desired response characteristic. The number of physical configurations which will yield a given result can be determined by transforming the matrix. However, now that their work, supplemented by additional solutions achieved elsewhere [9], has revealed the number of possible solutions obtainable with a wide variety of configurations, it could be expected that new and simpler solutions of the problem would emerge. That this has not occurred is unfortunate for the general solution of waveguide filters of this type; the constraints involved in order to obtain a solution involve lumped-circuit approximations and frequency-invariant couplings — approximations which restrict the band-

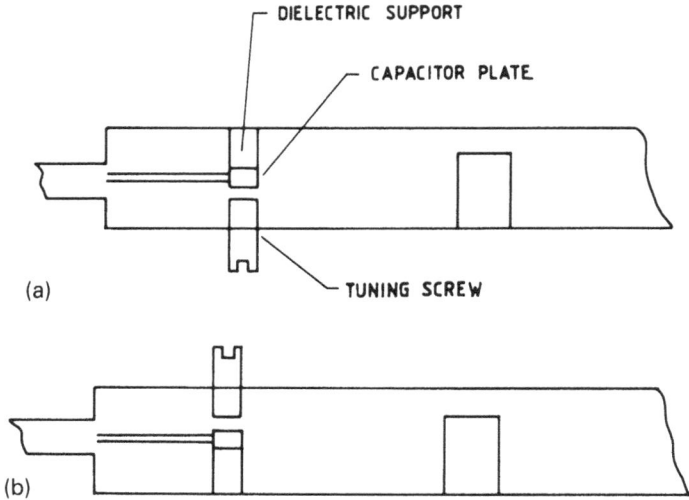

Fig. 4.30   (a) Coupling adjustment; (b) phase reversal.

widths that can be achieved with these filters to considerably less than those imposed by the restrictions of the configuration. Thus the very considerable advance that these filters represent has so far been restricted to narrow-band filters. It is probable that the time has come for a new approach to the problem: work done by the present authors indicates that with only minor modifications the circuit for these filters can be reduced to a lattice structure. With this simplification the solution of the bisected network in terms of its component diagonal and series arms becomes possible. Such an approach would make broader-band solutions practicable for the conventional filter and would also make a general comprehensive theory possible. With these suggestions for further work we now consider the theory in its present form.

4.6.1   Design theory

In making as much use of the existing theory as possible it is helpful to summarize the various steps in the theory (Fig. 4.32). The modified theory is illustrated in Fig. 4.33(a) and illustrates the similarities between the two approaches. First, we choose a given specification written in the frequency plane (Fig. 4.33(b)). By using eqn (3.1) which relates the lowpass and bandpass circuits we can write the lowpass frequency F expressed in terms of the cutoff frequency f_c, the center frequency f_0 and the frequency variable f as

$$F = \frac{f}{f_0} - \left\{ \frac{(f_c/f)^2 - 1}{(f_c/f_0)^2 - 1} \right\}^{1/2}$$

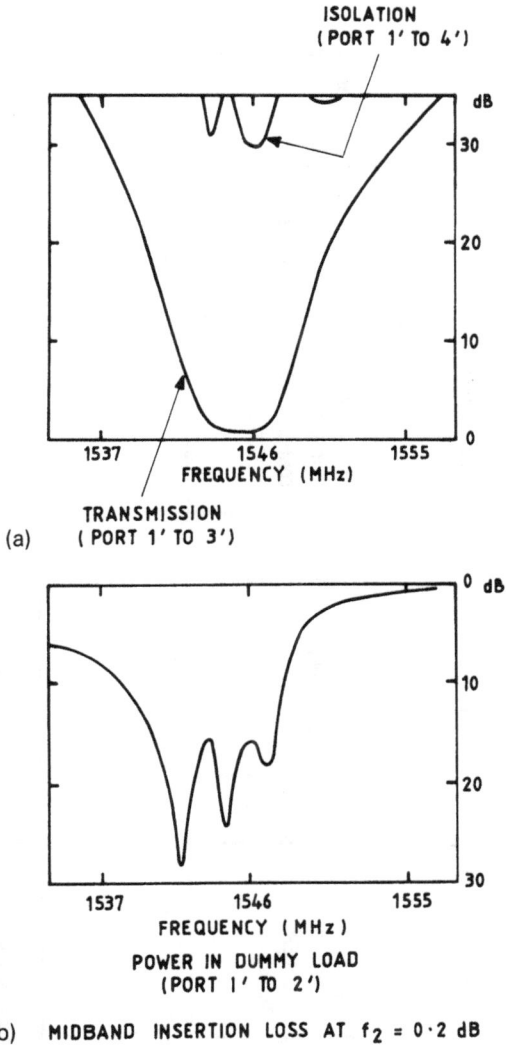

Fig. 4.31 Diplexer performance.

This gives us a response in terms of the lowpass frequency variable F as shown in Fig. 4.33(c). The next stage is to choose a realizable response and an appropriate configuration that will have sufficient resonators and couplings to achieve this response. Experience with the corresponding filters of the original type will be of value in carrying out this stage. The M-coupled filter is then designed using the program derived for the Atia and Williams filter. Resonators coupled by mutual inductance exactly correspond to the equivalent T-section for the

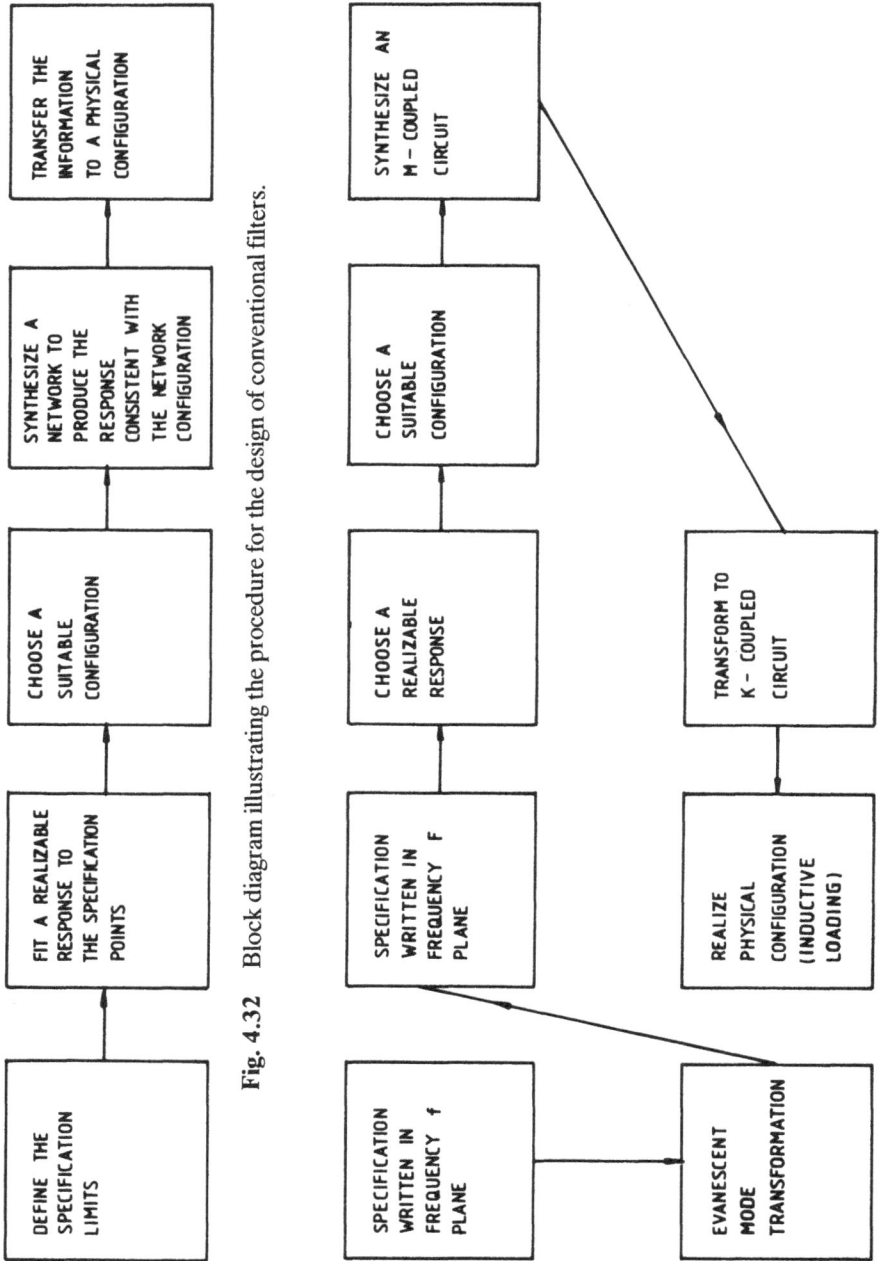

Fig. 4.32 Block diagram illustrating the procedure for the design of conventional filters.

Fig. 4.33(a)

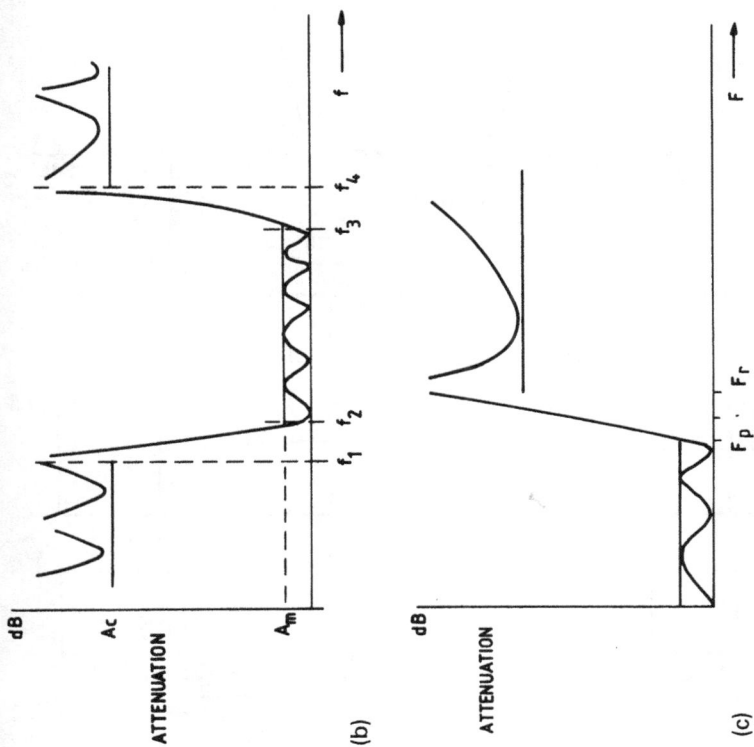

Fig. 4.33 (a) Block diagram illustrating the modified procedure for the design of evanescent mode filters; (b) specification in the frequency f plane; (c) specification in the new F plane.

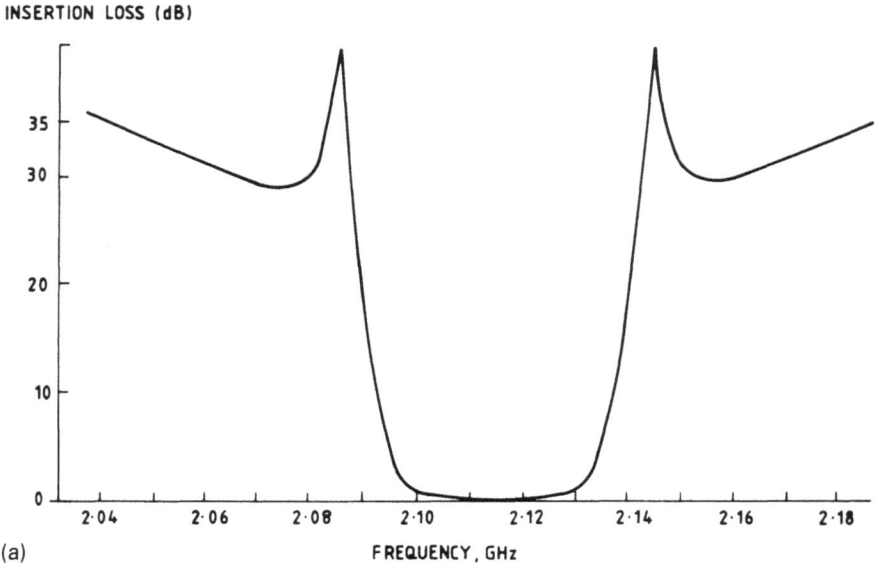

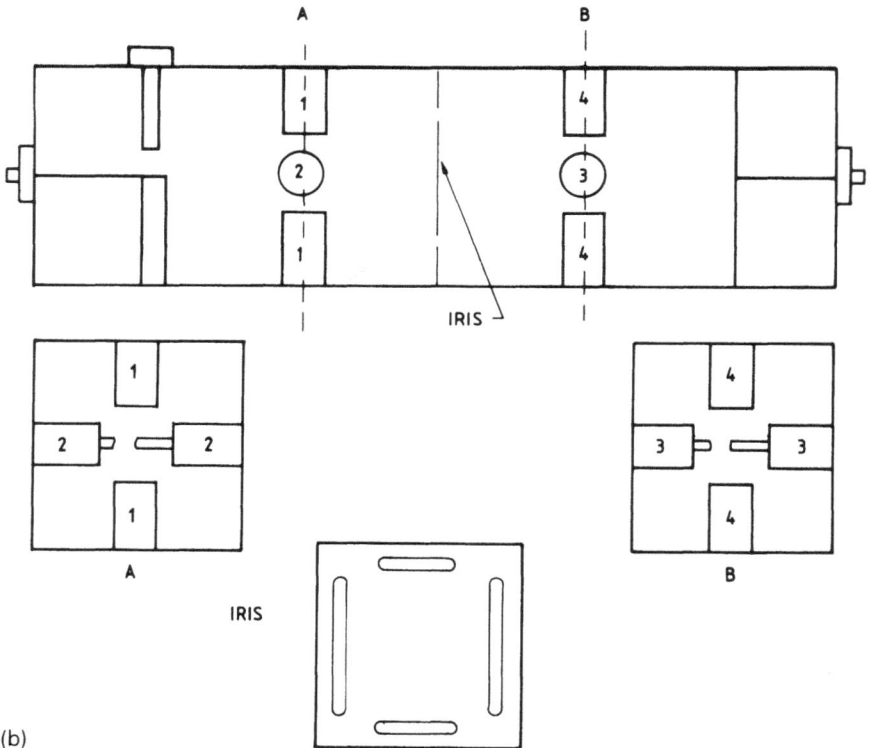

Fig. 4.34 (a) Evanescent mode elliptic function filter performance (insertion loss, 0.4 dB midband); (b) elliptic function filter configuration.

evanescent mode coupled filter. The elements of the equivalent π-section network follow directly from the T-section.

The final step in realizing the physical model involves inductive loading of the individual coupled resonators. Since each half of a dual-mode cavity has to be of identical length to the corresponding other half, inductive loading is employed to equalize these cavity lengths which will not ordinarily be identical. This is achieved using the methods outlined in Section 4.2.1.

4.6.2 Performance

The performance of an experimental model is shown in Fig. 4.34(a) and the construction is illustrated in Fig. 4.34(b). The model was of a comparatively elementary type involving only four resonators. In this sense it is the below-cutoff version which is similar to the original filter described by Williams [10], having a frequency of high rejection on each side of the passband. The loss of the filter was considered satisfactory bearing in mind its experimental construction in a brass waveguide. Because balanced capacitive posts are used to tune the resonators a small unbalance between the posts can be employed to introduce the cross-coupling between cavities that is an essential element in the operation of these filters. This is illustrated in Fig. 4.34(b), but it is extremely critical and is not recommended in more complex filters. The method of tuning which is suggested for more complex filters is a square slab of quartz which fills, or almost fills, the guide cross section. The cross-coupling can then be achieved with conventional schemes in which the coupling screw is mounted on the appropriate diagonal; a small amount of the quartz is ground away in order to make way for the screw. A more ambitious model included six dual-mode cavities (12 resonators) designed to produce a substantial degree of group-delay equalization plus a rejection frequency on each side of the passband. This experimental model was the approximate analog of the more conventional above-cutoff design outlined in a previous paper [4]. Some difficulty was experienced in tuning the filter for simultaneous realization of the rejection frequencies consistent with a ripple-free group-delay characteristic. It is believed that this was due to the highly critical nature of the scheme shown in Fig. 4.34(b). Quartz dielectric of suitable dimensions was not available at the time of these experiments.

References

1 Atia, A. and Williams, A. E. New types of waveguide bandpass filters for satellite transponders, *Comsat Tech. Rev.*, **1**, 21, Fall 1971.
2 Mok, C. K., Stopp, D. W. and Craven, G. Susceptance-loaded evanescent-mode waveguide filters, *Proc. Inst. Electr. Eng.*, **119** (4), 416, April 1972.

3 Mok, C. K. Design of evanescent mode diplexers, *IEEE Trans. Microwave Theory Tech.*, **MTT-21**, 43, January 1973.
4 Craven, G. and Radcliffe, C. Microwave filters for communication satellites, *Inst. Electr. Eng. Microwaves Opt. Acoust.*, **2** (5), 167, September 1978.
5 Lewis, W. D. and Tillotson, L. C. A non-reflecting branching filter for microwaves, *Bell Syst. Tech. J.*, **27**, 83, 1948.
6 Ragan, G. L. *MIT Radiation Laboratory Series*, Vol. 9, Section 10.10, McGraw-Hill, New York, 1948.
7 Cohn, S. B. and Coale, F. S. Directional channel-separation filters, *Proc. IRE*, **44**, 1018, August 1956.
8 Matthaei, G. L., Young, L. and Jones, E. M. T. *Microwave Filters, Impedance-matching Networks and Coupling Structures*, Chapter 14, McGraw-Hill, New York, 1964.
9 Pfitzenmaier, G. Synthesis and realization of narrow-band canonical microwave bandpass filters exhibiting linear phase and transmission zeros, *IEEE Trans. Microwave Theory Tech.*, **MTT-30** (9), 1300, September 1982.
10 Williams, A. E. A four-cavity elliptic waveguide filter, *IEEE Trans. Microwave Theory Tech.*, **MTT-18** (12), 1109, December 1970.

Chapter 5

Additional Components

5.1 Introduction

Although filters represent the most complex components used in a micro-wave subsystem, a number of additional components are required in a complete unit. One of the most important of these is the interface between the passive elements and an active or nonlinear device, often referred to as the crystal mount. In addition to this requirement a number of passive elements, such as hybrid-Ts, directional couplers, phase shifters, transitions to other units, impedance transformers and bends, are frequently needed. Many of these components are considered in this section. Nonreciprocal components are omitted for the moment but are discussed later in a separate chapter. Firstly we consider diode mounts.

5.2 Diode mounts

Arguably, the most obvious technique is to employ a waveguide transformer as represented by the crossbar transition illustrated in Fig. 5.1 [1]. This

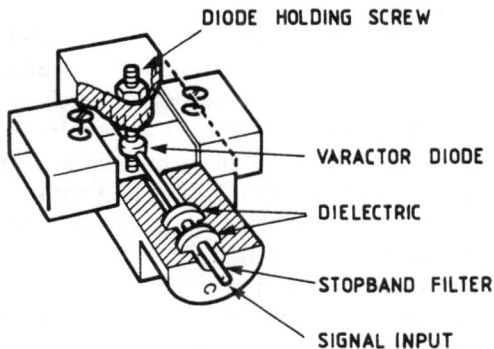

Fig. 5.1   Varactor diode mount.

85

method has been used successfully with varactor devices. However, the purpose of the crossbar is open to misinterpretation. The intention is twofold: to reduce the lead length to the varactor and so to reduce the inductance in series with the active device, and to reduce the impedance (both real and imaginary parts) that the device presents at its output terminals. Matching is then easier and is finally achieved by appropriate choice of the distance between the device and its nearest resonator. For example, if the previous resonator is the last element of a filter, that network will have been designed to operate into a load that is normalized to the absolute magnitude of the waveguide admittance. The coupling length must then be chosen to provide a match between the active device and the impedance located at the adjacent resonator. This of course necessitates a knowledge of the normalized admittance of the active device, which will have to be measured at the appropriate level that would occur in conventional practice. The method of measurement is different, however. In this case we determine the bandwidth of the device, considered as a separate unit, when excited at the prescribed power level. The load impedance can be obtained from the 3 dB bandwidth and the distance necessary to match the two impedances can be determined. The technique will become clearer after the discussion of impedance matching transformers later in this chapter. A second version of a crossbar transformer is shown in Fig. 5.2 [2]. Its construction is similar to that shown in Fig. 5.1 except for the stub projecting from the guide. In addition, the stub facilitates resonances at the harmonically related frequencies at the input and output frequencies of a harmonic multiplier. This is considered in greater detail in Chapter 8.

5.2.1   Image-terminated mixers

In many applications, especially mainline microwave links, it is important that the input match be very good and remain so despite later replacement of the input mixer diode whilst the unit is in service. This requirement has led to the widespread use of isolators between the signal input filter and the mixer in order to maintain the desired input match. The disadvantage which results from this practice is that power at the image frequency is dissipated in the input isolator and a potential source of improvement in the overall noise figure of the receiver is lost. The improvement in noise figure which can be achieved with reactive termination of the image has been described by Johnson [3] who discusses both open-circuit and short-circuit termination of the image. Although the greatest improvement in noise figure is obtained with open-circuit termination, this type is more difficult to achieve; short-circuit termination is easier to realize in practice. A mixer which embodies short-circuit termination of both the image and the sum frequencies is shown in Fig. 5.3.

As is well known [4] power is generated at the image frequency by the second harmonic of the local oscillator; the mechanism of the generated image is clear from the following. If we call the local oscillator frequency f_o, the signal

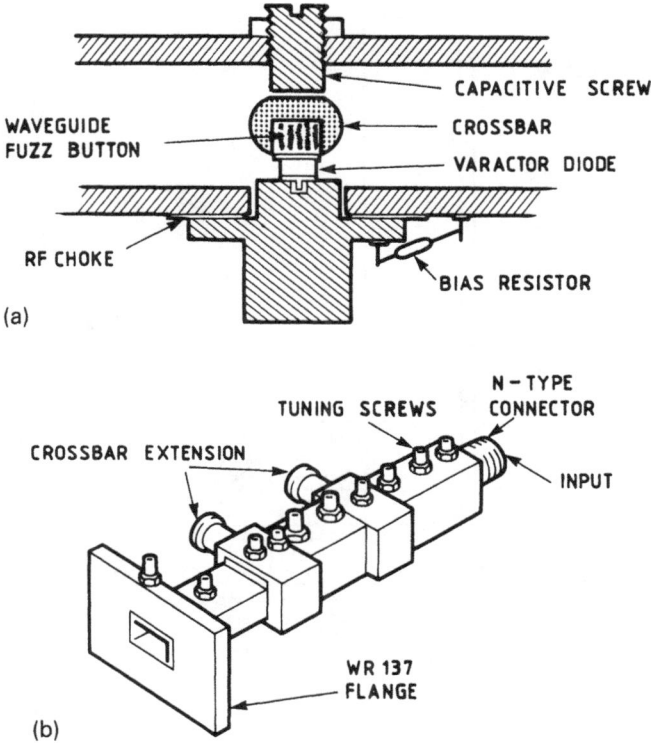

CAPACITIVE SCREW

WAVEGUIDE
FUZZ BUTTON

CROSSBAR

VARACTOR DIODE

RF CHOKE

BIAS RESISTOR

(a)

N-TYPE
CONNECTOR

TUNING SCREWS

CROSSBAR EXTENSION

INPUT

WR 137
FLANGE

(b)

Fig. 5.2 Frequency multiplier: (a) section through the diode mount; (b) view of the multiplier.

frequency f_s and the intermediate frequency f_i, the primary intermediate frequency is obtained from

$$f_i = f_s - f_o \tag{5.1}$$

However, at the harmonic $2f_o$ we have

$$f_i = 2f_o - (f_s + f_o) \tag{5.2}$$

The scheme shown in Fig. 5.3 is based on the longitudinal slot obstacle described in Section 2.11. The loading on the resonator is controlled by both the slot length and the point at which the diode is located on the slot length. The slot primarily tunes the resonator, and the tuning screw is employed only for exact adjustment of the resonant frequency. The short-circuit termination of the image is achieved by the screw located in the evanescent guide which covers the slot. By the usual law of impedance inversion a parallel r sonance occurring at this plane will appear as a short-circuit across the diode The bandwidth of the

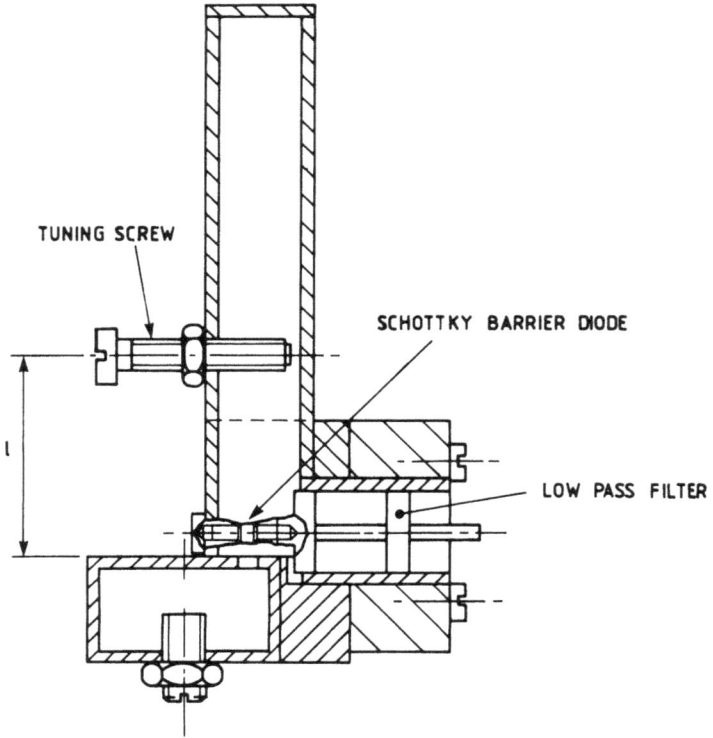

Fig. 5.3   Image-terminated mixer mount.

effective short-circuit is controlled by the coupling length l. If l is too short then the band over which high reflection exists will extend into the signal frequency band; too great a value of l will result in a narrow bandwidth at the image frequency and will lead to group-delay problems in the mixer. Termination of the sum frequency is achieved by the position of the short-circuit backing plate which closes off the guide. At the sum frequency the guide is propagating freely and so the length of the guide can be used in this way to obtain the correct termination of the sum frequency. The group-delay characteristics of the mixer were checked against a standard waveguide mixer and were found to be completely adequate. An important feature in achieving this result was a broad-band filter of the type described earlier which prevented the sum frequency travelling into the main filter. This is not shown in the diagram. The overall improvement in noise figure obtained by terminating both sum and image frequencies was 1.3 dB [5], a figure which was confirmed by direct comparison. With some further refinement of the lowpass filter configuration, which was needed to reduce the conversion loss in the absence of image termination, the performance would have been comparable with the best obtainable. Unfortunately, the mixer met the desired commercial specifications as it stood, and time was not available for this extra work. The unit is discussed further in Chapter 8.

5.3 Impedance transformers

The impedance inversion properties of a section of cutoff guide were discussed in Chapter 3. Located between two resonators it effectively behaves like a section of quarter-wave line. By choosing the impedance of the "quarter-wave" line to be the geometric mean between two different load resistances, the latter can be matched to each other by a similar technique to that employed in classical transmission line theory. In Chapter 4 it was shown that the impedance transformation obtained in a given physical length is increased by inductive loading. It is therefore reasonable to expect that capacitive loading will have a corresponding effect but in the opposite direction. Consequently we can employ a variable-capacitance obstacle — a capacitive screw — as a means of adjusting the electrical length of a given physical length of cutoff guide which is between two resonators. This represents a useful variable transformer which permits empirical matching between two impedance levels to be effected quite simply. All that is necessary is to make the physical length rather greater than is necessary to match the expected loads. The capacitive screw is then introduced into the guide and the electrical length of the guide is reduced until matching of the source is achieved.

5.3.1 Multisection transformers

This approach can be extended to multisection transformers. For example, Bharj and Mak [6] have employed the method to design a transition from propagating waveguide to microstrip which is much cheaper than an orthodox ridge guide or exponential taper. In addition, it avoids the problem of dc blocking which otherwise is often necessary with the former transitions. The principal objective was to provide a simple cheap transition for a specific application. Its construction and performance are shown in Figs 5.4(a) and 5.4(b). Wide-band performance was not needed in the application, but in general where a broader-band perormance is necessary the susceptance existing at the junction would be a disadvantage. The capacitive tuning screws are attractive from an economic viewpoint, but dielectric capacitors are desirable where close spacing between resonators is essential for broader-band performance.

The computed performance of a broader-band version of a transformer of similar basic design but intended only to match impedances located in a cutoff guide is shown in Fig. 5.5. The design theory can be simply summarized. First, decide on the number of transformer sections to be employed. Next, select from

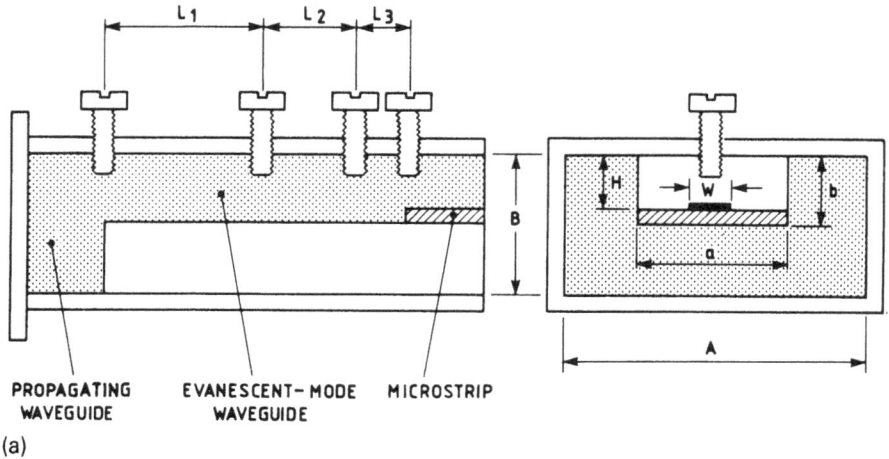

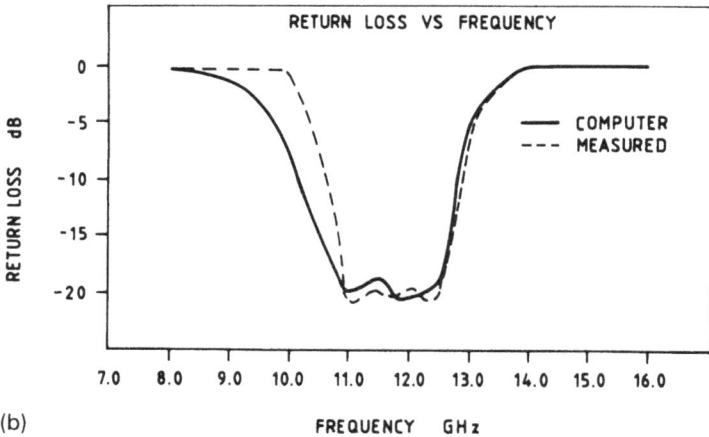

Fig. 5.4   Transition from propagating waveguide to microstrip using a cutoff waveguide: (a) construction; (b) performance.

the table of Matthaei *et al.* [7] the normalized impedance ratios that would be
required with corresponding quarter-wave transformers in order to match the
desired load resistances. These ratios represent the values of the normalized
impedances (the $\sinh(\alpha l)$ values) throughout the filter. The distances between
resonators can then be determined from the known cutoff frequency of the
guide. Where broad-band transformers are essential the obstacles employed are
best represented by dielectric strips; the reasons for this choice are discussed in
Chapter 3. Since the foregoing approach adapts a design theory for quarter-
wave line transformers to what, after all, are impedance-transforming filters, it is
not necessarily the ideal theory. One of the problems encountered in the design

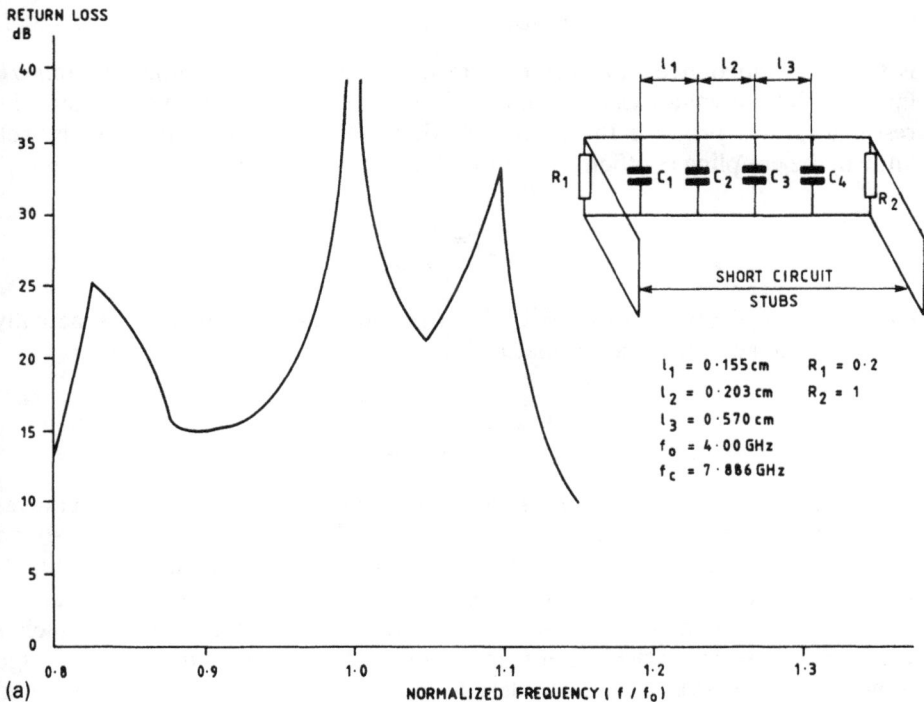

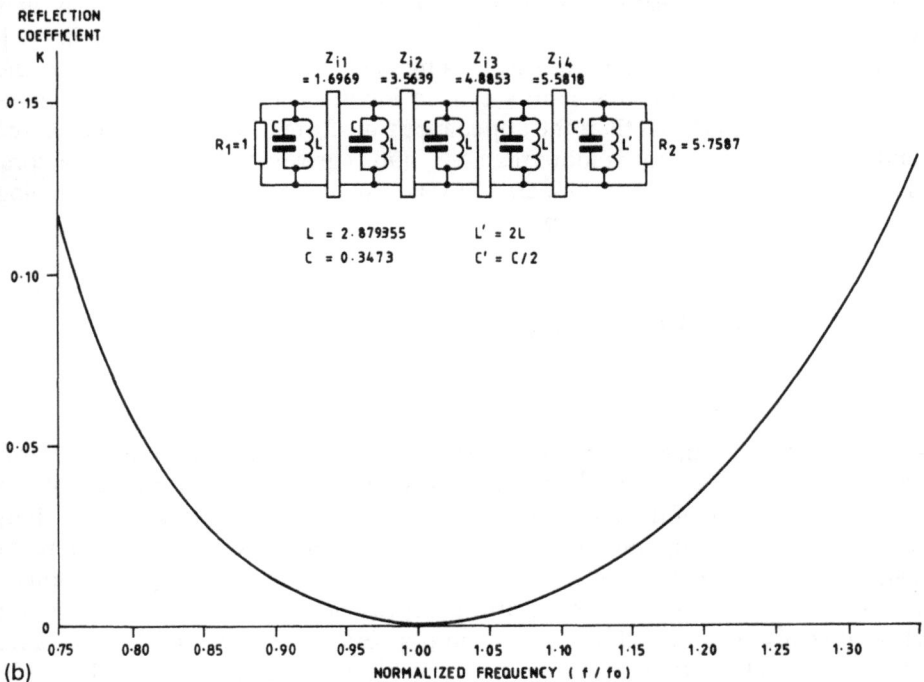

Fig. 5.5 (a) Wider-band design based on stepped transformer theory; (b) impedance transformer formed by a bisected nine-resonator filter.

is that the impedance transformers are determined by the values of $\sinh(\alpha l)$ involved. These values yield the equivalent loading resistances which shunt the resonators and produce the doubly loaded Q of each resonator. As is well known, the coupling coefficient required for optimally coupled resonators is

$$K_{\text{OPT}} = \frac{1}{Q_1 Q_2}$$

However, it is shown elsewhere [8] that the coupling coefficient which actually exists between cutoff guide resonators is given by

$$K = \frac{1}{\cosh(\alpha l)}$$

and is, of course, already determined by the value of αl required in the equivalent "quarter-wave" transformer. This introduces little difficulty in narrow-band designs where $\sinh(\alpha l)$ and $\cosh(\alpha l)$ are virtually identical, but at broader bandwidths the two values approach the limiting values of zero and unity respectively. This makes the realization of a given design more difficult, a fact which is largely responsible for the higher ripple values achieved in the theoretical design shown in Fig. 5.5(a). A different design approach would seem to be indicated here, and in this respect it is worth considering a theory based wholly on filter design. In coupled-resonator filters the bandwidth which is necessary in each individual resonator is achieved by raising the impedance level at which the resonator operates; the resonators can be identical. For example, in a nine-resonator maximally flat filter the impedance transformation between the source and the center resonator is more than 5:1, and is greater with more resonators. The performance of a filter bisected in this way and terminated in a matching impedance is shown in Fig. 5.5(b).

5.4 Hybrid-Ts and directional couplers

5.4.1 Hybrid-T

A hybrid-T (magic-T) is illustrated in Fig. 5.6 [8]. As in its above-cutoff forerunner the isolation between the series and shunt arms is a result of the intrinsic symmetry of the construction. The equivalent circuit in Fig. 5.6(b) illustrates the mode of operation. The series arm is represented by a parallel resonant circuit located at the junction and was referred to in an early publication as a π-section resonator. The shunt arm is represented by a resonator located a short distance away which transforms into a series circuit at the junction. The lumped equivalent circuit is readily recognizable as a bridged-T circuit, derivable as a special case of the lattice structure. The model was an early one and such

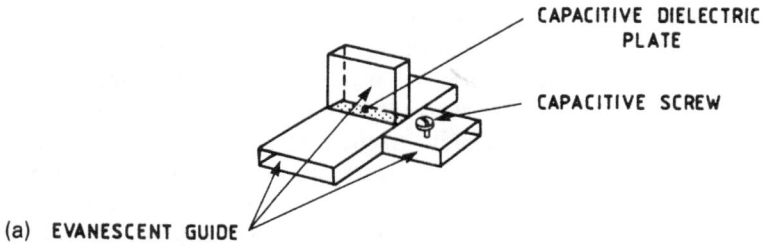

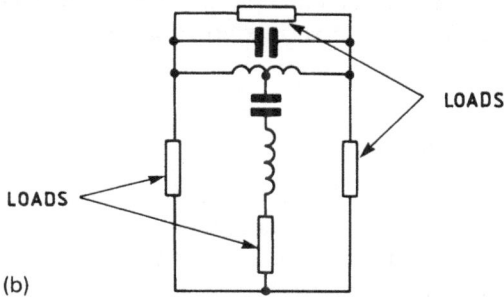

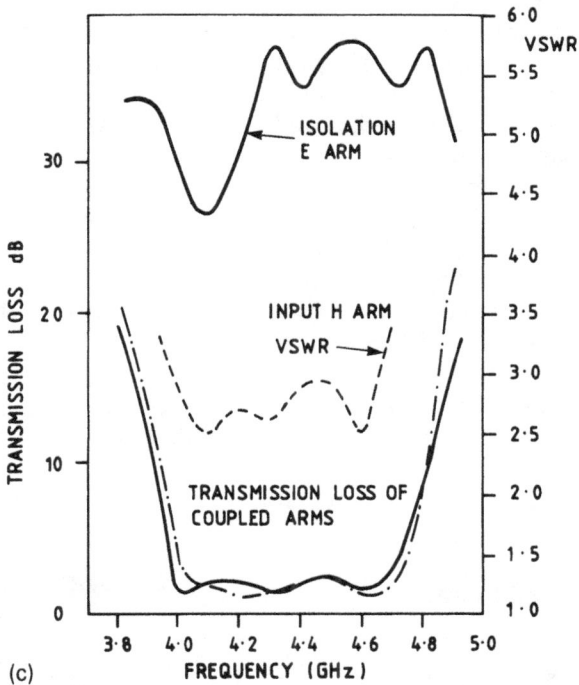

Fig. 5.6 Hybrid-T using a cutoff waveguide: (a) construction; (b) equivalent circuit; (c) performance. (Reprinted with permission of the *Microwave Journal*, from the August 1970 issue. © 1970 Horizon House — Microwave Inc.)

shortcomings as it has result from the matching techniques in use at the time. A more recent version has been constructed by Begemann and Peters [9].

It will be clear that we can arrive at the following components by the limiting process of introducing short-circuits at various planes. For example, if we introduce a short-circuit at the plane of the dielectric sheet, the junction reduces to a shunt-connected T-junction; the corresponding series T-junction requires the short-circuit to be loaded at the side wall where the two guides meet. If we go a step further and imagine that one wall on the shunt arm is extended across the guide whilst a short-circuit is located at the series arm junction, then the junction reduces to a right-angle bend. Most junctions of this type have been built and tested.

5.4.2 Directional couplers

The concept of directional couplers is so closely associated with freely propagating waves that the notion of directional couplers employing resonators in a cutoff guide may seem at first sight something of a contradiction. However, investigation in detail shows that the contradiction is more apparent than real.

Directional couplers have been considered in terms of their properties as symmetrical four-port networks [10, 11]. For example, Reed and Wheeler [10] show that such a network can be bisected along the line of symmetry as shown in Fig. 5.7. The product matrix of the combined elements for the open-circuit and short-circuit conditions (Figs 5.8(a) and 5.8(b) respectively) then enables the outputs at the four ports to be obtained. The magnitudes of A_1 and A_4 are found separately from the sum and difference respectively of (half) the reflection coefficients computed from the open-circuit and short-circuit matrices. Similarly, the A_2 and A_3 terms are obtained from the sum and difference respectively of (half) the insertion voltage transmission between matched loads in the foregoing matrices. The method of analysis is particularly useful in the treatment of branched-guide couplers and hybrid rings.

Inspection of the equivalent circuit of coupled resonators in a cutoff guide (Fig. 5.9(a)) reveals the similarities between the two circuits. The J-inverters replace the quarter-wave lines and have essentially the same properties, but

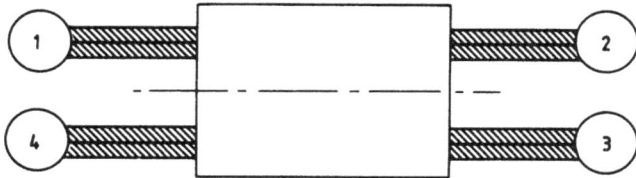

Fig. 5.7 Bisected directional coupler.

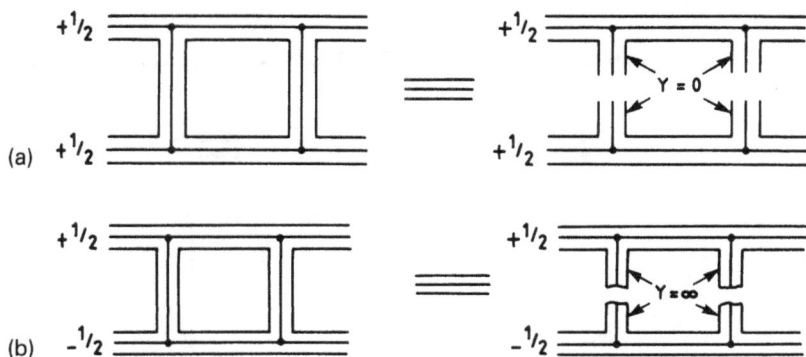

Fig. 5.8 Circuits used in the analysis of a directional coupler: (a) open-circuit condition; (b) short-circuit condition.

unlike the latter element the phase constant remains frequency invariant and the magnitude of the characteristic (image) impedance is a function of frequency. In addition, we have the resonator which terminates the inverter at each end. The consequence of bisecting such a network along the line of symmetry in Fig. 5.9(a) is that we have two "half-inverters", i.e. two networks of electrical length $\pi/4$ with frequency-dependent impedance but frequency-invariant phase.

The realization of the directional coupler is straightforward and is depicted in Fig. 5.9(b). For small coupling a hole in the common side wall between the two guides is sufficient. The coupling between the resonators in the main and auxiliary guides can be regarded in a similar way to the inductively loaded filters discussed in Chapter 4. In each case the inductive loading reduces the J-inverter admittance (equivalent to an increased value of $\sinh(\alpha l)$). Experimental models of this type have been built and tested and have yielded satisfactory performance, but experimental details are not available. Instead, computed performance curves are displayed in Fig. 5.10.

5.5 Transitions to other media

Transitions to coaxial cable have been described in Section 3.3 which also includes some discussion of a suitable measurement technique for the experimental development of loops. A loop which has been employed in broaderband coupling is described elsewhere in the text (Section 7.5). Capacitive probes located in the high-electric-field region have also been used in some applications and have the merit of extreme simplicity (Fig. 5.11(a)). One form of this technique, which should be suited to wide-band resonators, is shown in Fig. 5.11(b).

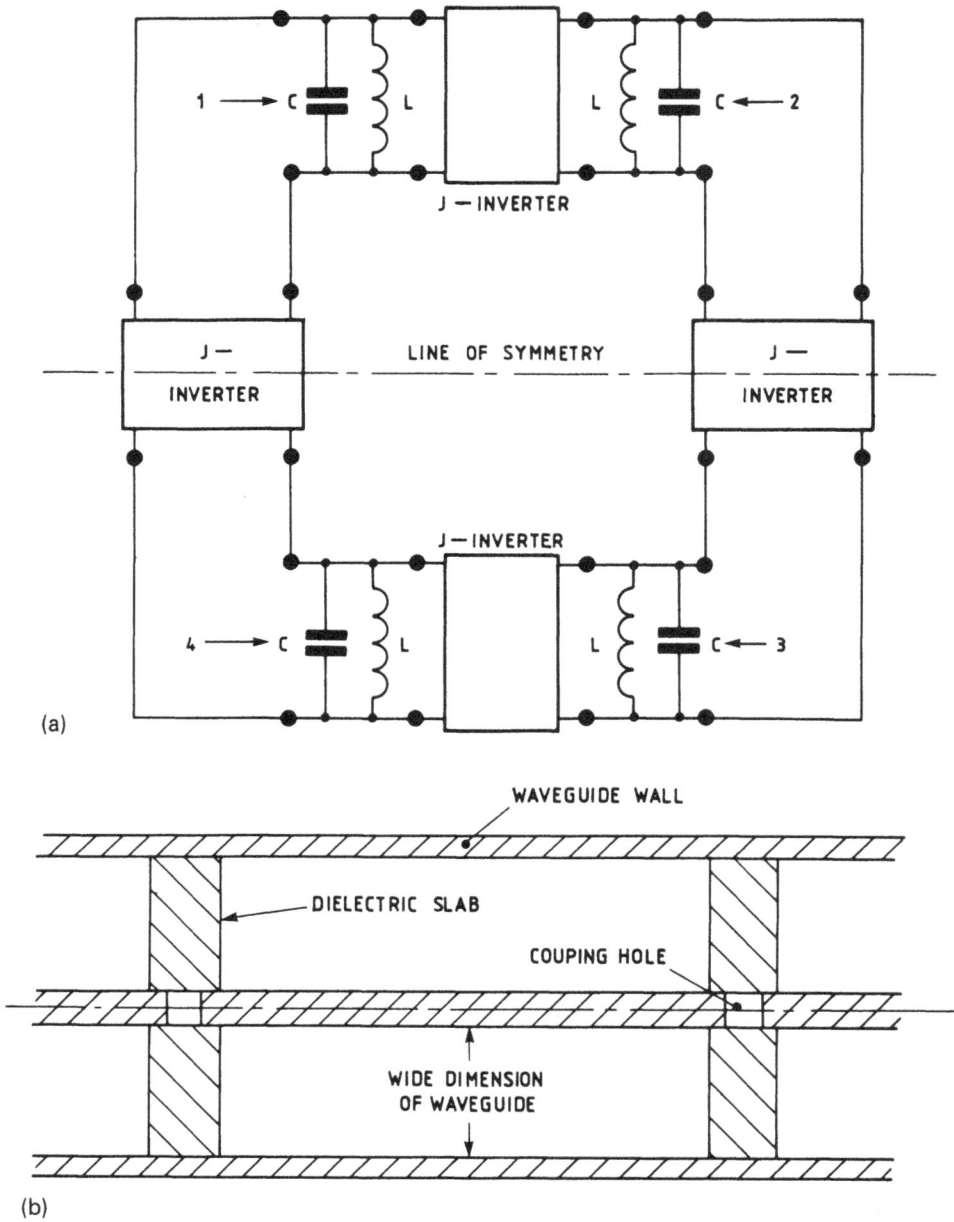

(a)

(b)

Fig. 5.9   Narrow-band directional coupler: (a) equivalent circuit; (b) sectional view of the construction.

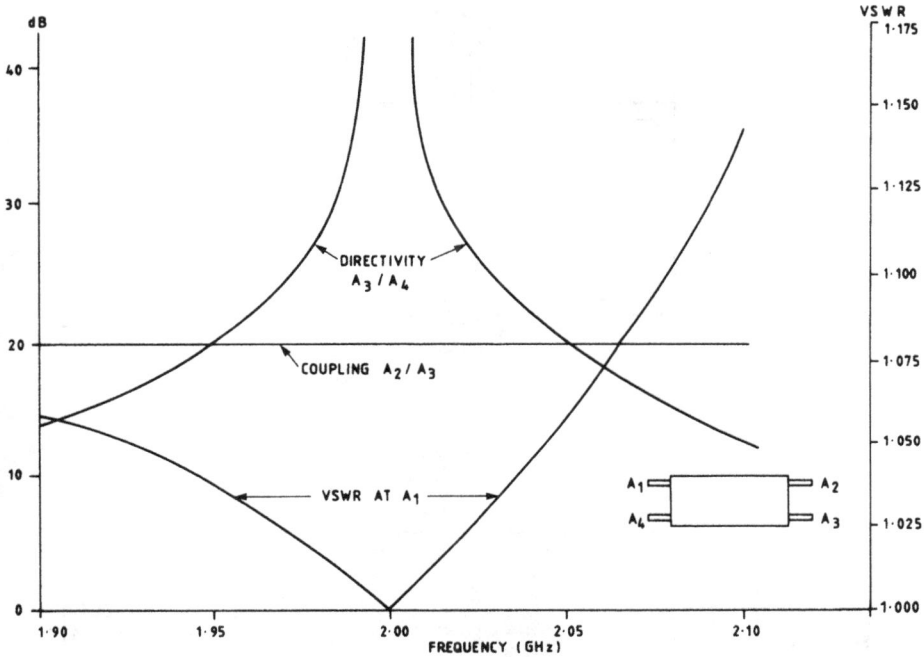

Fig. 5.10 Computed performance of a narrow-band 20 dB directional coupler.

Mention should also be made of the use of two- or three-element lowpass filters to transform the impedance of the input circuit to that of the filter [12].

Transitions from cutoff guide resonators to microstrip have already been described, but transitions from propagating waveguide to cutoff guide remain to be considered. Ordinarily, this will produce an inductive junction susceptance unless the propagating guide is of the same dimensions but loaded continuously with dielectric. The net junction susceptance has been considered previously [13] as the algebraic sum of the capacitative susceptance, which results from the different height dimensions, and the inductive susceptance, which is caused by the different guide widths. However, this derivation is approximate, and an expression derived by Lewin [14] should be more accurate. Tapers in the broad side between the two guides, whether continuous or stepped, are not of course acceptable because of the proximity to cutoff which is encountered at some stage in the taper. However, tapers which employ ridged guide to connect the two media can be used to eliminate the junction effect which results from the different dimensions. (Of course the smaller junction effect resulting at the end of the ridge will have to be taken into account.)

Components are always developed as separate units, and one of the most common methods of connecting them is by coaxial cable. This is helpful in

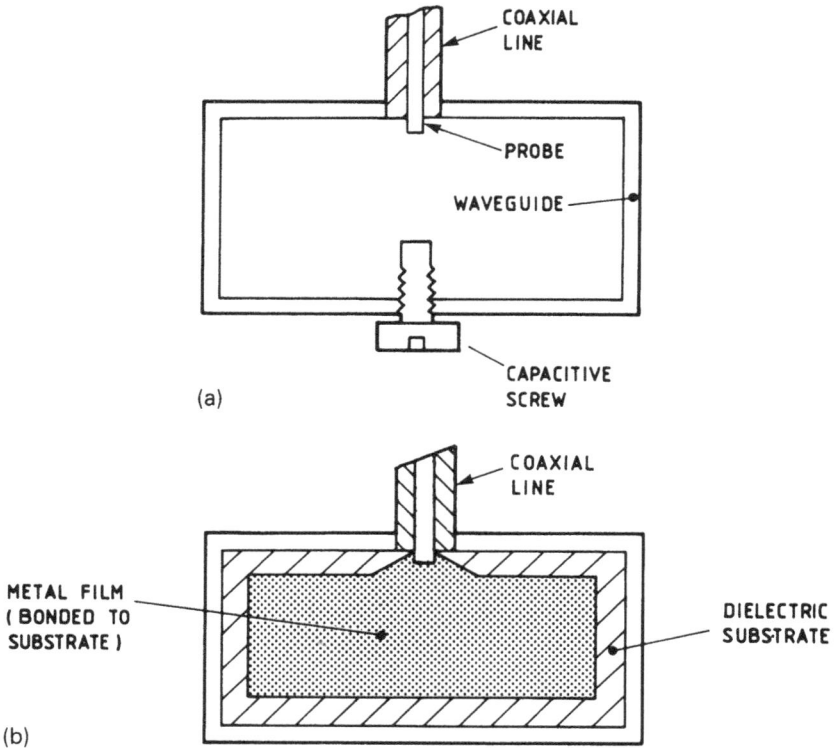

Fig. 5.11   Methods of capacitive coupling: (a) narrow band; (b) broad band.

major systems because the specifications required in the individual components are calculated by the system designers; the performance can then be measured separately against these specifications. This approach to system design also confers a welcome degree of flexibility on the physical disposition of the components. At the same time it introduces the undesired complexity of additional connectors.

Evanescent mode components can be tested separately as waveguide components via a suitable waveguide flange, which is inherently simpler and more reliable, although bulkier, than a corresponding coaxial connector. The technique is to design the component to match a specified real impedance, ordinarily normalized to unity, and then to design the other components to the same impedance level. Of course this method only works if all components are constructed in the same guide size and no junction susceptances are created at the flanges. Otherwise there are no special problems involved. It is clear that no additional lengths of guide should be included. Minor problems in testing waveguide components to be employed in this way are discussed in Chapter 9.

References

1 Kwiatkowski, W., Arthanayake, T. and Knight, V. H. Efficient high-level up-converter, *Electron. Lett.*, **6** (19), 625, 17 September 1970.

2 Dahele, J. S. and Hill, D. R. Slimguide frequency multipliers for microwave applications, *Electr. Commun.*, **49** (1), 65, 1974.

3 Johnson, K. M. X-band integrated circuit mixer with reactively terminated image, *IEEE Trans. Microwave Theory Tech.*, **MTT-16** (7), 388, July 1968.

4 Torrey, H. C. and Whitmer, C. A. *Crystal Rectifiers*, MIT Radiation Laboratory Series, Vol. 15, McGraw-Hill, New York, 1948.

5 Craven, G. Slimguide microwave components, *Electr. Commun.*, **47** (4), 245, 1972.

6 Bharj, S. S. and Mak, S. Waveguide-to-microstrip transition uses evanescent mode, *Microwaves Radiofreq.*, 99, January 1984.

7 Matthaei, G. L., Young, L. and Jones, E. M. T. *Microwave Filters, Impedance-matching Networks and Coupling Structures*, Section 6, McGraw-Hill, New York, 1964.

8 Craven, G. Waveguide below cutoff: a new type of microwave integrated circuit, *Microwave J.*, **13** (8), 51, August 1970.

9 Begemann, G. and Peters, C. X-band balanced mixer using evanescent mode circuitry, *Electron. Lett.*, **14** (24), 748, 23 November 1978.

10 Reed, J. and Wheeler, G. J. A method of analysis of symmetrical four-port networks, *Trans. IRE*, **MTT-4**, 246, October 1956.

11 Young, L. Branch guide couplers, *Proc. Natl Electronics Conf.*, **12**, 723, 1956.

12 Snyder, R. V. New application of evanescent mode waveguide to filter design, *IEEE Trans. Microwave Theory Tech.*, **25** (12), 1013, December 1977.

13 Craven, G. F. and Mok, C. K. The design of evanescent mode filters for a prescribed insertion loss characteristic, *IEEE Trans. Microwave Theory Tech.*, **MTT-19** (3), 295, March 1971.

14 Lewin, L. *Theory of Waveguides*, Chapter 8, p. 249, Newnes-Butterworths, London, 1975.

Chapter 6

Ferrite-loaded Devices in Waveguides Below Cutoff

6.1 Introduction

The invention of microwave ferrite devices in the early 1950s revolutionized some aspects of the current technique by introducing the principle of nonreciprocal behavior into a range of microwave components. Since then it has become mandatory that any passive network circuitry which seeks to be competitive with existing techniques must also provide this facility. Our initial concern was with the creation of a range of passive reciprocal devices in waveguide below cutoff and the idea that nonreciprocal devices — especially circulators and isolators — were also possible was not immediately obvious. This situation was changed by the invention of the junction circulator in cutoff guides [1] in which the relevant property is the ferrite resonant cavity. However, the approach to nonreciprocal devices was more gradual than this might imply because this work was preceded by earlier theoretical and practical investigations in which ferrite, and its applied magnetic field, was employed to control the behavior of purely reciprocal devices [2–5]. Since these devices took precedence historically, it seems logical to deal with them first.

6.2 Magnetically tunable bandpass filter

The propagation characteristics of electromagnetic waves in ferrite-loaded waveguides have been studied in detail by a number of workers [6–8]. For the ferrite-loaded guide in Fig. 6.1 it has been shown that the propagation characteristics of, for example, the TE_{10} mode are dependent on the applied magnetic field. Of particular relevance to our present interest, the cutoff frequency can be varied by the dc magnetic field about the value that would apply if the ferrite were simply a dielectric of the same permittivity. This result is a direct consequence of the dependence of the rf permeability of the ferrite on the magnetic field intensity. This dependence can be exploited in several devices of

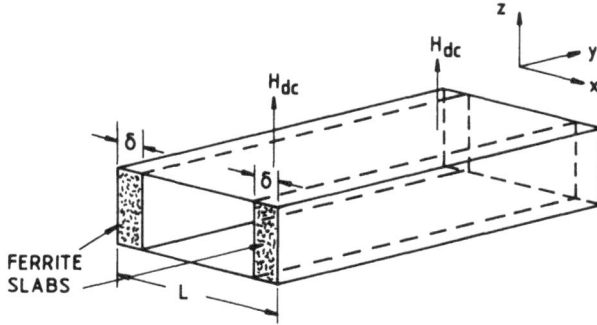

Fig. 6.1   Twin ferrite slabs at the side walls of a rectangular waveguide.

which the most obvious is a magnetically tunable filter. The idea first emerged some time after publication of the original letter disclosing the principles of the evanescent mode filter [9].

The scheme was described initially in a brief letter [2] and later in a more detailed paper [3]. The intention was to provide a filter which achieved precise tuning of a number of resonators over a moderate band rather than to realize wide-band tuning of simpler filters. The scheme is illustrated in Fig. 6.2(a).

6.2.1   Design theory

In the following discussion it is assumed that the ferrite-loaded guide shown in Fig. 6.1 is infinitely long and that there is no variation in the rf field in the direction of the dc magnetic field. With these simplifications the modes that are considered to exist in the region are limited to the TE_{n0} class. Specifically, we are interested in the TE_{10} mode below its cutoff frequency.

By specifying that the rf field components of these modes vary as $\exp\{j(k_a x - \beta y)\}$ and $\exp(-K_m x - j\beta y)$ in the empty and ferrite-filled regions respectively, it follows, on assuming the usual $\exp(j\omega t)$ time dependence, that

$$k_a^2 = \omega^2 \mu_0 \varepsilon_0 - \beta^2 \tag{6.1}$$

$$K_m^2 = -\omega^2 \mu_e \varepsilon + \beta^2 \tag{6.2}$$

where $\mu_e = \mu_0(\mu^2 - \kappa^2)/\mu$ is the scalar rf effective permeability of the ferrite, $\varepsilon = \varepsilon_r \varepsilon_0$ is the rf permittivity of the ferrite and

$$\bar{\bar{\mu}} = \mu_0 \begin{pmatrix} \mu & -j\kappa & 0 \\ j\kappa & \mu & 0 \\ 0 & 0 & 1 \end{pmatrix}$$

is the intrinsic rf permeability tensor of the ferrite. If the ferrite is assumed to be lossless the elements of the rf permeability tensor can be conveniently expressed

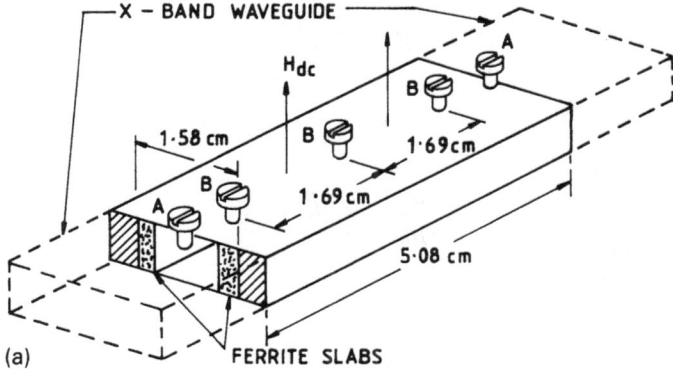

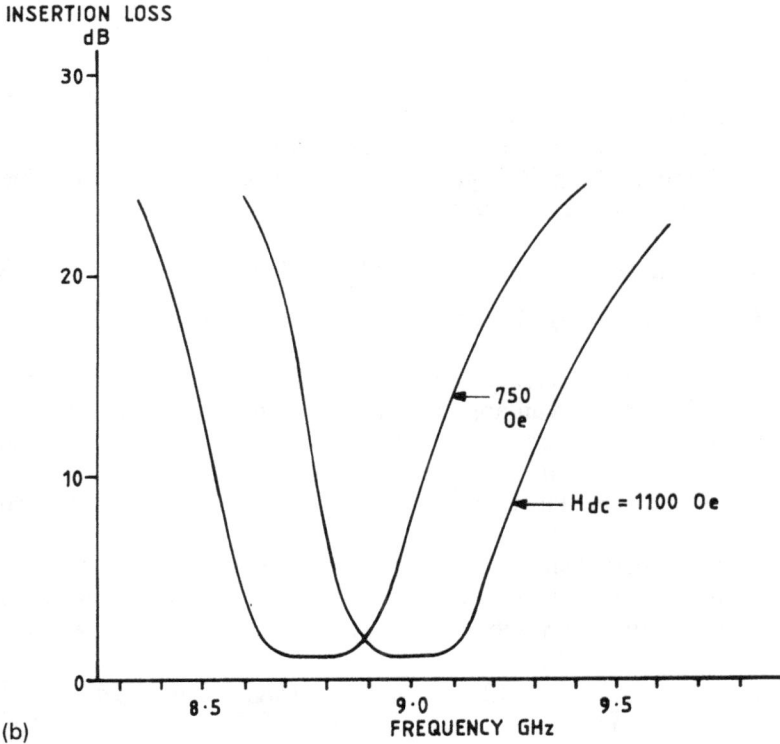

Fig. 6.2 Magnetically tunable three-section bandpass filter using a ferrite-loaded evanescent mode waveguide: (a) filter construction showing the ferrite slabs against the waveguide side walls; (b) performance of the filter showing the tuning action of the magnetic field.

in terms of the usual parameters p and σ as follows:

$$\mu = 1 - \frac{p\sigma}{1 - \sigma^2} \qquad (6.3)$$

$$\kappa = \frac{p}{1 - \sigma^2} \qquad (6.4)$$

$$\frac{\mu_e}{\mu_0} = \frac{1 - (p + \sigma)^2}{1 - p\sigma - \sigma^2} \qquad (6.5)$$

where

$$p = \frac{\Gamma}{\omega} 4\pi M_s$$

$$\sigma = \frac{\Gamma}{\omega} H_i$$

H_i is the internal dc magnetic field within the ferrite, $\Gamma = 2.8$ MHz Oe^{-1} is the gyromagnetic ratio and $4\pi M_s$ is the saturation magnetization of the ferrite.

Consideration of eqn (6.2) gives the following conditions for producing an evanescent state for the TE_{10} mode, i.e. $\beta^2 < 0$.

Case 1: K_m real, $\mu_e/\mu_0 < 0$ and $|\omega^2 \mu_e \varepsilon| > |K_m^2|$. Hence $\beta = -j\alpha$.

Case 2: K_m imaginary, $\mu_e/\mu_0 > 0$ and $|\omega^2 \mu_e \varepsilon| < |K_m^2|$. Hence $\beta = -j\alpha$.

In each case the necessary conditions can be obtained by adjusting the dc magnetic field strength to an appropriate value. Thus, in the first case μ_e/μ_0 is made negative and as large as required, whereas in the second case μ_e/μ_0 is made positive and as small as is necessary. The dependence of μ_e/μ_0 on the parameters p and σ is illustrated in Fig. 6.3; these curves were derived for a lossless ferrite.

The ferrite-loaded structure of Fig. 6.1 has been analyzed (refs. 10 and 11) and the characteristic equation for the propagation constant β was obtained by formulating the boundary value problem. From ref. 11, Chapter 9, we obtain

$$\tan\{k_a(L - 2\delta)\} = \frac{(2\mu_0 K_m/\mu_e k_a) \cosh(K_m\delta) \sinh(K_m\delta)}{\{1 - (\mu_0\beta/\mu_e k_a \theta)^2\} \sinh^2(K_m\delta) - (\mu_0 K_m/\mu_e k_a)^2 \cosh^2(K_m\delta)}$$

$$(6.6)$$

where $\theta = j\mu/\kappa$. It is apparent from eqn (6.6) that the propagation constant β occurs only as an even power and consequently the ferrite-loaded guide is reciprocal as regards β. Since the equation is transcendental an appropriate

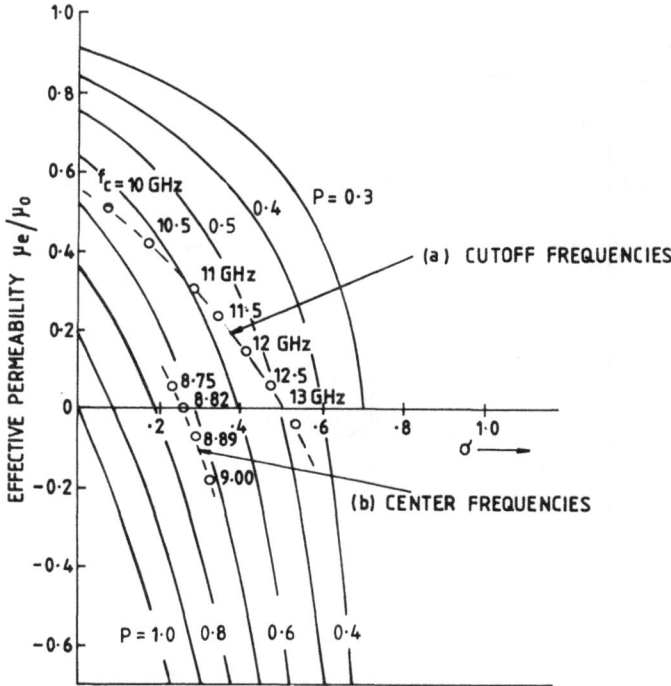

Fig. 6.3 Effective permeability μ_e/μ_0 as a function of α with p as a parameter: – – –, curves for the evanescent mode filter of Fig. 6.2 showing the variation of μ_e/μ_0 with (a) the cutoff frequency and (b) the center frequency.

computer program is necessary for its solution. To obtain the cutoff frequency we substitute $\beta = 0$ into (6.6) and, after some rearrangement, we have

$$\frac{\mu_e \varepsilon_0}{\mu_0 \varepsilon} \tan\{\omega_c(\mu_e \varepsilon)^{1/2}\delta\} = \cot\left\{\omega_c(\mu_0 \varepsilon_0)^{1/2} \frac{L - 2\delta}{2}\right\} \qquad (6.7)$$

where ω_c is the cutoff angular frequency.

Some results obtained by previous workers (ref. 11, p. 385 and Fig. 9.33) are reproduced in Fig. 6.4, curve A, which shows the variation in phase constant β as a function of the thickness of the ferrite slab. The inset depicts the electric field pattern in the waveguide. Modes of this type have been designated modified TE_{10}. We now consider the design of this type of filter.

6.2.2 Filter design principles

The initial work on these filters was carried out at an early stage in the development program when the only design method available was based on

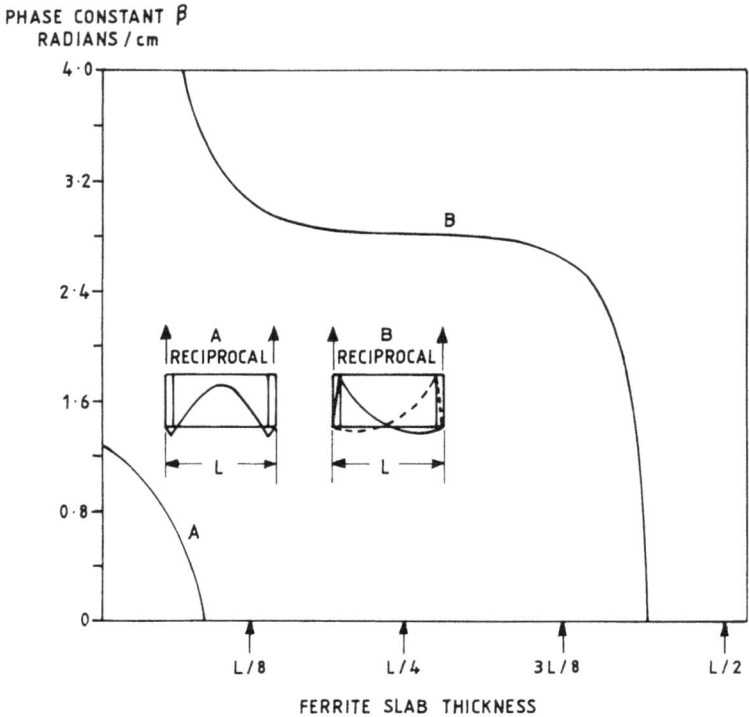

Fig. 6.4 Variation of the phase constant as a function of the thickness of the ferrite slab. The inset shows the electric field patterns for the TE_{10} mode (curve A) and the ferrite-dielectric mode (curve B).

image parameter theory [12]. Although this theory yields a less desirable passband response than more sophisticated approaches, it is well suited to a filter which embodies a uniform ferrite loading intended to tune a number of identical resonators.

One of the first decisions that has to be made is whether μ_e/μ_0 is negative or positive (case 1 or case 2 referred to above). A negative μ_e/μ_0 should exclude more of the rf field energy from the ferrite than a positive μ_e/μ_0. When the linewidth of the ferrite is comparatively broad a negative μ_e/μ_0 seems to be the more reasonable choice on the grounds that it will reduce the ferrite loss. Whether this conclusion is also true with narrow-linewidth ferrites remains to be investigated, but in any case the increased loss in the ferrite that occurs as gyromagnetic resonance is approached with increasing magnetic field also needs to be considered. Obviously, the choice is complex. Some of these problems are illustrated in the experimental design described below.

6.2.3 Experimental filter

The filter is shown in Fig. 6.2(a) and includes a ferrite section at each side wall ($l = 5.08$ cm, $h = 1.02$ cm and $\delta \approx 0.20$ cm). The ferrite was mounted in a WR62 waveguide and the filter was tested by coupling it to a WR90 guide; its linewidth ΔH was 150 Oe and its relative permittivity ε_r was 12. The saturation magnetization $4\pi M_s$ was 2300 G.

Initially, the capacitive screws were withdrawn from the guide and the experimental cutoff frequency of the structure was measured as a function of the applied magnetic field. The results in Fig. 6.5 show the agreement between theory and experiment. The internal field in the ferrite shown in the figure was derived from the applied field by making due allowance for the demagnetizing effect of a structure of this geometry. Also of interest in view of the earlier comments is the variation in μ_e/μ_0 as a function of cutoff frequency. These results are shown in Fig. 6.3 as a broken line. It will be seen that μ_e/μ_0 has a positive value of 0.55 at 9.7 GHz, passes through zero at 12.8 GHz and has a small negative value of about 0.05 at the cutoff frequency 13 GHz.

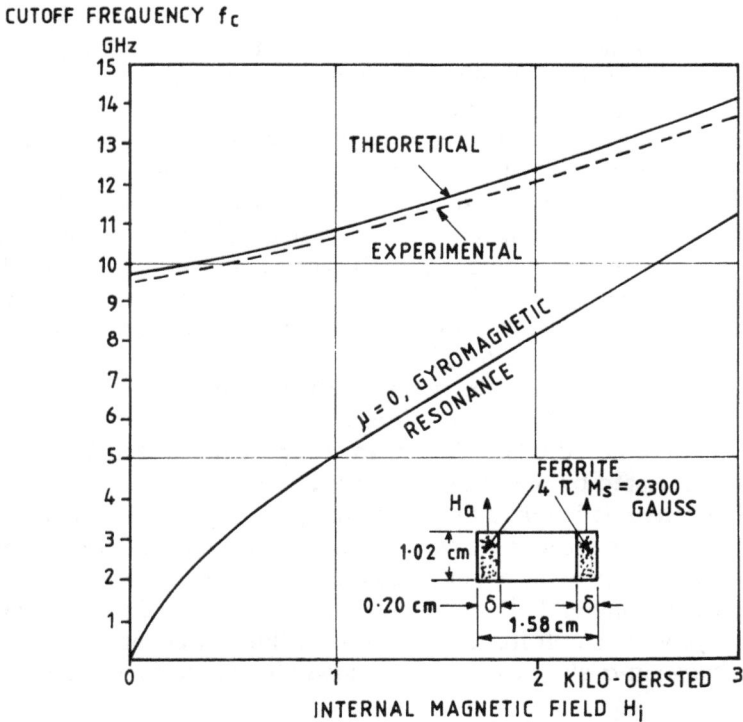

Fig. 6.5 Cutoff frequency in a ferrite-loaded waveguide as a function of the internal magnetic field H_i.

Once the experimental confirmation of the theoretical predictions concerning cutoff frequency was obtained, this information was used to set the cutoff frequency to 11.25 GHz. This nominal design frequency corresponded to a dc magnetic field strength of about 1000 Oe. The filter was then tuned to resonate at 9.0 GHz using the method based on the input poles and zeros of reactance [13], a technique which is necessary because of the high insertion loss (about 100 dB) of the cutoff guide in its untuned state. The technique is well known, and in its simplest form consists of tuning for input admittance values which are alternately located diametrically on the circumference (unit reflection coefficient) of the Smith chart. Where the filter is connected to a propagating guide this will necessitate tuning out the junction susceptance. If it is assumed that the propagating guide admittance (real) is terminated in a nonpropagating guide of equal numerical (but imaginary) value, the correct admittance presented by the cutoff guide when the junction susceptance is tuned out will be $Y = -j1$. Therefore, if we initially short-circuit the junction and adjust the reference plane accordingly, the next step is to alter the tuning screw penetration until the admittance presented on the Smith chart is the desired value. If the two guide admittances are not of the same absolute magnitude, an appropriate point on the chart must be chosen.

The performance of the filter is shown in Fig. 6.2(b). Apart from the increased insertion loss resulting from the presence of the ferrite, the performance is typical of such a filter [12]. The use of one of the ferrites that are currently available with a much narrower linewidth would greatly improve the loss. The magnetic tuning of the center frequency has been limited to 300 MHz; greater values lead to an excessive change in the overall bandwidth. The reason for this and ways of reducing it are discussed in detail later. The measured bandwidth at the center of its tuning range is 4% compared with a calculated value of 6%. The discrepancy is attributed to a simplification in the design which involved using an approximate value for the attenuation coefficient α instead of the accurate solution of eqn (6.6).

The variation in the ferrite permeability over the tuning range is shown by the broken line in Fig. 6.3: μ_e/μ_0 varies from a positive value of 0.062 at 8.75 GHz to -0.178 at 9.0 GHz. Clearly, the range is much less than the possible maximum indicated by the complete curve and it is reasonable to enquire into the cause of the restriction in the tuning range. This lies in the change in bandwidth which becomes excessive when the filter is tuned over a wider range and is basic to a design involving a continuous ferrite throughout the filter. It is worth recalling from Chapter 1 the principles on which filters of this type are based. Whilst varying the cutoff frequency tunes the filter by changing α and thereby Z_0, the change in α also changes the bandwidth since to first order the latter is given by

$$\frac{f_2 - f_1}{f_0} = \frac{2}{\sinh(\alpha l)} \tag{6.8}$$

If we are to overcome this problem we must adopt a similar approach to that employed in Section 4.2 in which we defined two regions: the resonator region and the coupling region. By allowing the ferrite to encroach somewhat into the coupling region, we can make the increase in α which results from magnetic tuning offset the natural decrease which occurs in an unloaded region as the frequency increases. The approach [4] is illustrated in Fig. 6.6 in which the tuning performance of a continuously loaded ferrite-tuned filter (broken curve) is compared with that of a unit in which the ferrite occupies only 50% of the total filter length (full curve). The improved results in terms of constancy of bandwidth are self-evident.

Partial ferrite loading brings additional possibilities. The polycrystalline ferrite can be replaced by single high-Q crystals so that much lower losses can be achieved. Since the early work was completed the increased use of computers has made more sophisticated designs attainable. It is possible to depart from the initial image parameter concepts and realize prescribed insertion loss properties [14]. For example, Snyder [15] has employed these principles in a five-resonator structure using stepped slabs of ferrite and has achieved promising results: the tuning range is from 8.6 to 9.2 GHz, corresponding to a dc magnetic field varied between 1000 and 1600 Oe; the 1 dB bandwidth is in the range of 100 MHz with an insertion loss of only 0.75 dB. The unit is illustrated in Fig. 6.7.

The magnetically tunable filter in a cutoff guide is intended to complement the familiar yttrium iron garnet (YIG) filter by providing a low-loss multisection unit which can be tuned over a moderate frequency band. Remote tuning of filters offering broader bandwidths than were previously achievable and with prescribed transmission characteristics is now possible.

6.3   T–R switch [5]

An additional application for the magnetically tunable filter described in the previous section is in a type of T–R switch [5]. It is illustrated in Fig. 6.8 and comprises a length of guide which is divided by an H-plane bifurcation into two reduced-height waveguides: one is operated below cutoff and the other above cutoff. The former represents a filter of the type already described and the latter is a section of conventional propagating waveguide. The two states of the switch can best be described by considering the off state first. The filter is tuned to the center of its passband by the applied magnetic field and so the input wave divides between the two paths. The two paths are arranged so that at their outputs the waves are of equal magnitude but opposite phase. Thus complete cancellation occurs and total reflection results.

In the on state the filter is detuned by reducing the magnetic field on the ferrite-tuned filter and the junction is matched such that complete transmission

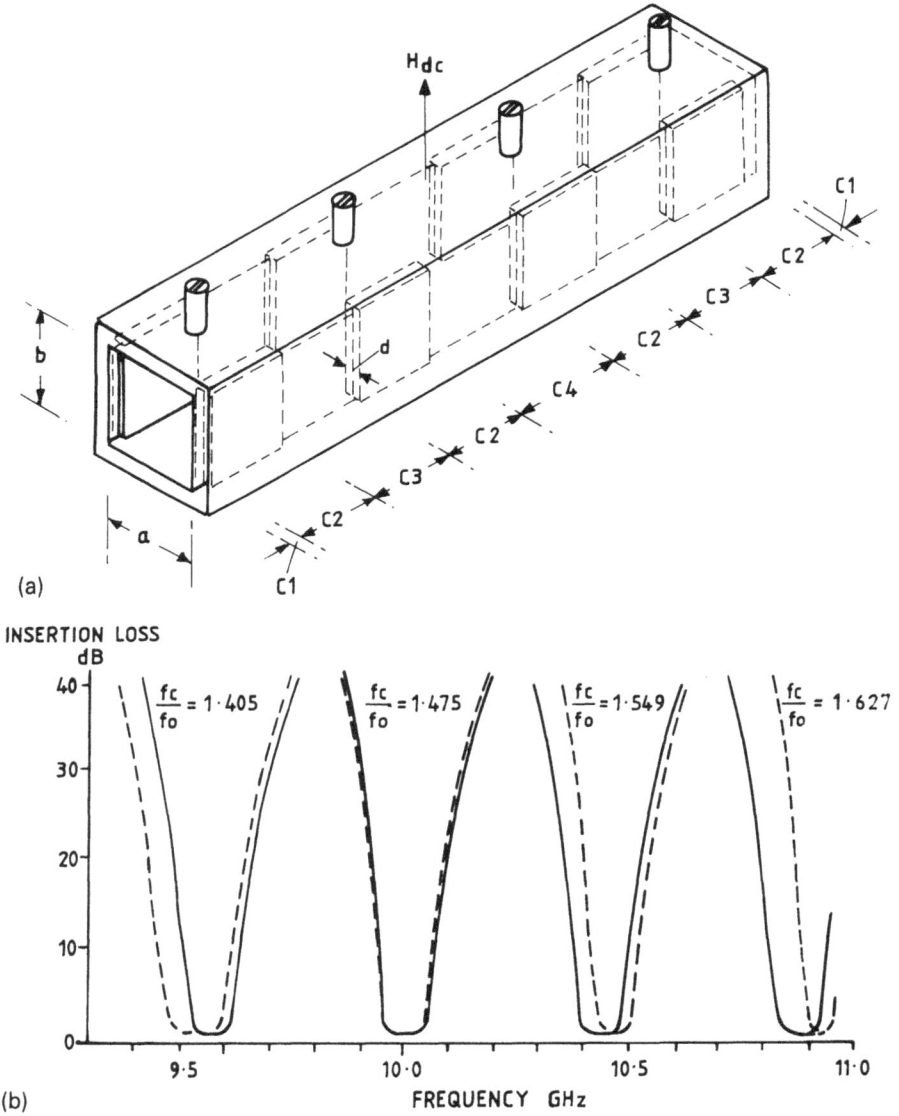

Fig. 6.6 Magnetically tunable four-section bandpass filter using a partially-ferrite-loaded evanescent mode waveguide: (a) filter construction; (b) performance of the filter (——, ferrite loading 50% of the total length; – – –, ferrite loading 100% of the total length).

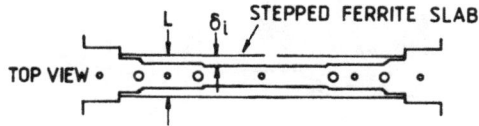

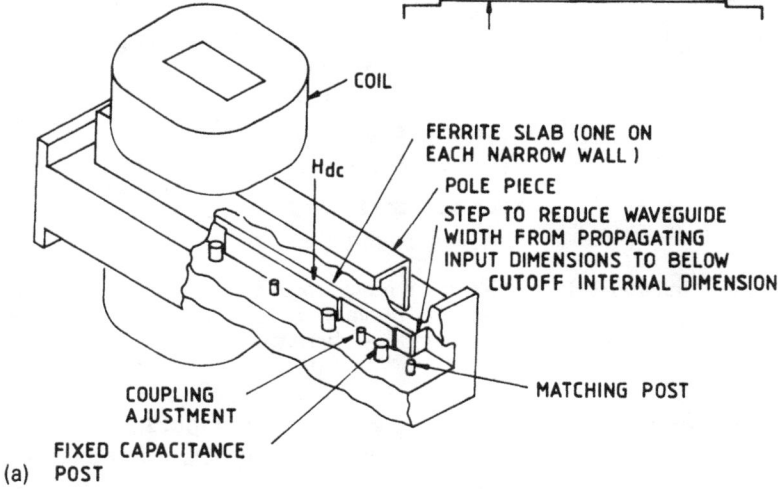

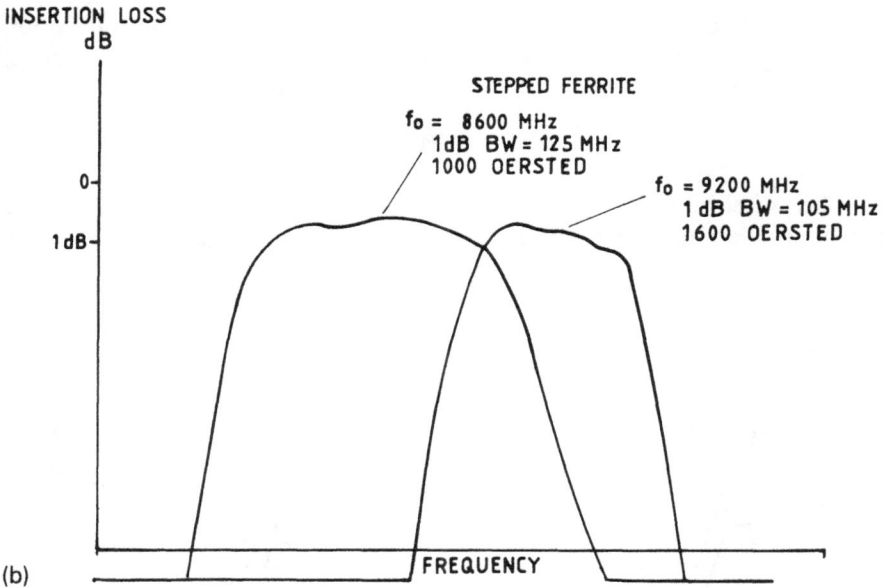

Fig. 6.7 Magnetically tunable five-pole bandpass filter using stepped ferrite slabs in an evanescent mode waveguide: (a) filter construction; (b) performance of the filter.

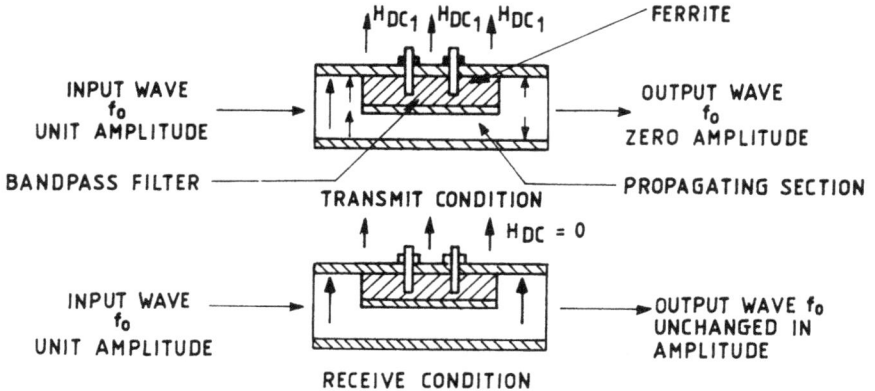

Fig. 6.8 Experimental T–R switch using a ferrite-loaded evanescent mode waveguide. (Reprinted with permission of the *Microwave Journal*, from the August 1970 issue. © 1970 Horizon House — Microwave Inc.)

occurs. The importance of the matching screws at the input to the filter is that they provide an independent means of controlling the input impedance of the filter. This facilitates good rejection in the OFF position of the switch. The phase length of each resonator section at midband is $\pi/2$ radians, independent of the physical length. Consequently, an n-section filter will have a phase shift of $n\pi/2$ rad. It then follows that the phase length of the reduced-height guide should also be an integral number of $\pi/2$ rad (say m), and for cancellation

$$|n - m| = 2(2r + 1) \tag{6.9}$$

where $r = 0, 1, 2$ etc.

The switch shown in Fig. 6.8 was constructed in a WR90 guide, with the metal bifurcation dividing the guides into two regions of equal height. The width

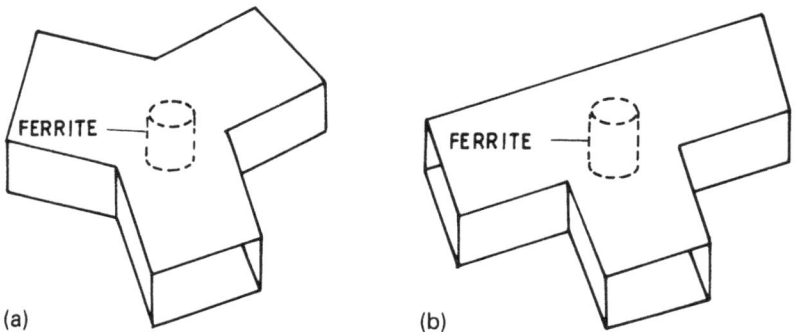

Fig. 6.9 Conventional three-port waveguide junction circulators: (a) Y-junction; (b) T-junction.

of the upper section was reduced to 1.58 cm, and the filter contained two ferrite slabs 0.203 cm thick. An isolation of 43 dB was obtained over a 30 MHz bandwidth when the filter was tuned to midband (magnetic field of 890 Oe). The insertion loss with the field at zero (ON position) was 0.3 dB.

These results were considered quite good for a feasibility design. Where greater rejection is required units can be cascaded. A wider rejection bandwidth would be desirable, and this could be achieved by employing a broader-band filter which would have a smaller phase variation over the band. An advantage of this type of switch in radar systems is that it can be independently triggered earlier than the transmitter pulse, thus eliminating the breakthrough of energy spikes. An important feature of such a switch is the operating speed, but this has not been investigated.

6.4   Circulators

The three-port circulator is one of the most important and widely used microwave components. It appears in most systems either with its basic circulator role or, by terminating one of its ports in a matched load, as an isolator. In practice the three-port circulator has been realized in most component media: lumped circuits, striplines, microstrips, coaxial lines and rectangular guides, including ridged guides. Operating frequencies range from 100 MHz to beyond 100 GHz. Thus such a component is essential if the technique using waveguide below its cutoff is to be truly comprehensive.

Conventional three-port waveguide circulators are generally constructed in standard-size waveguide. The waveguide sections are joined together to form a symmetrical Y- or T-junction as shown in Fig. 6.9, and the ferrite material is located centrally in the junction and magnetized by a suitable value of externally applied magnetic field. The ferrite-loaded junction forms a resonant cavity which is coupled to each of the output ports by a section of propagating guide operating in the dominant mode. The basic difference between this general arrangement and the present circulator lies in the coupling between this resonator, which is located in the reduced-width guide, and the external resonators that form the output ports. The first three-port circulator which was built [1] was constructed in WR62 guide (cutoff frequency, 9.49 GHz) and is shown in Fig. 6.10. The circulator was designed to work at around 8.5 GHz and, for convenience in initial testing, an interface with standard WR90 guide was used. At the time that this unit was developed existing designs of waveguide junction circulators were largely empirical, and therefore techniques that were current at that time were employed to obtain the desired performance. The ferrite resonator was located at the center of the junction and the magnetic field was varied in order to obtain the correct resonant frequency. In conjunction with this operation the coupling to the two-resonator filter shown in the diagram

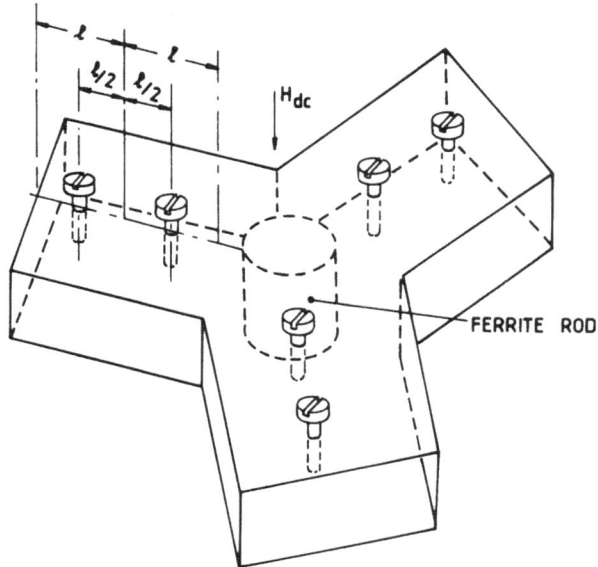

Fig. 6.10   Three-port junction circulator using a waveguide operated below the cutoff frequency of the TE_{10} mode.

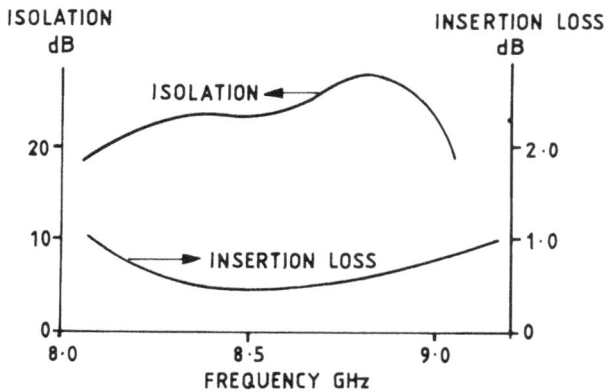

Fig. 6.11   Performance of an evanescent mode junction circulator.

was varied (the filter had been designed and tuned previously) until optimum performance was obtained. The results are shown in Fig. 6.11.

By the standards of present-day performance the results shown are comparatively modest; the principal objective was to demonstrate feasibility. However, they represented an important start which excited interest elsewhere

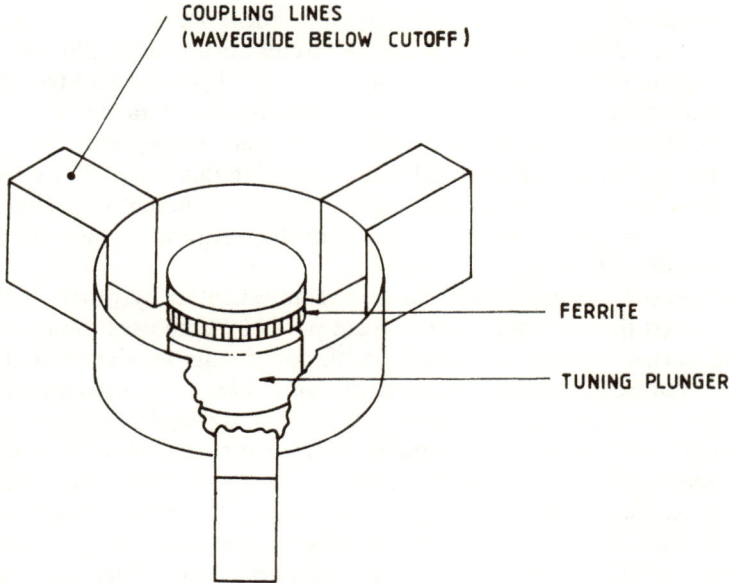

Fig. 6.12   Three-port circulator.

Table 6.1

Frequency range (GHz)	Isolation (dB)	Bandwidth (GHz)	Insertion loss (dB)	Temperature range (°C)
4	35	0.5	0.05	0 to + 40
11	30	0.9	0.1	− 20 to + 50
12	35	0.5	0.1	0 to + 40
14	25	0.5	0.1	− 40 to + 70
20	20	2	0.25	− 20 to + 50
30	20	2	0.35	− 20 to + 50
50	20	3	0.5	− 20 to + 50

and led to the development of units with excellent performance by other workers in the field. One of these units [16] is shown in Fig. 6.12. These workers reported that the weight of this type of circulator was considerably less than that of the corresponding propagating waveguide junction and that the performance was superior to an equivalent unit in microstrip. A useful technique in tuning the circulator is to terminate one arm in a variable short-circuit and make adjustments to the magnetic field whilst observing the behavior at the other two ports. Additionally in the experimental model (Fig. 6.12) a tuning plunger is used to obtain optimum results. The performance is shown in Table 6.1 together with that of several other circulators developed by these workers. It is worth noting that one of the advantages of this type of circulator lies in its filter-like characteristics: a good performance is obtained over a substantial bandwidth beyond which attenuation rises quite steeply. In certain circumstances this enables a designer to dispense with a filter, and a significant improvement in the noise figure of parametric amplifiers has been achieved in this way [17]. The comparatively modest bandwidths quoted in Table 6.1 are not indicative of limitations in performance, but merely that the system design required a high performance over that bandwidth only.

The question of bandwidth has been investigated by other workers [18] who concluded that two approaches were possible. One uses the basic method described earlier [1, 16] but differs in minor details as shown in Fig. 6.13. Schieblich and Schünemann described the operation of the circulator in terms of the equivalent circuit (Fig. 6.14) which was derived [18] after the initial invention [1] of this type of circulator. This approach uses the inverter and remnant inductances described in detail in Chapter 3. The condition for ideal circulation is then that the complex conjugate Z_c^* of the ferrite resonator impedance (and its load) should match the source impedance Z_0. The problem is discussed in greater detail in the original paper [18]. When matched to the source by a two-resonator filter the overall bandwidth over which the isolation exceeded 20 dB was 600 MHz about a center frequency of approximately 9.3

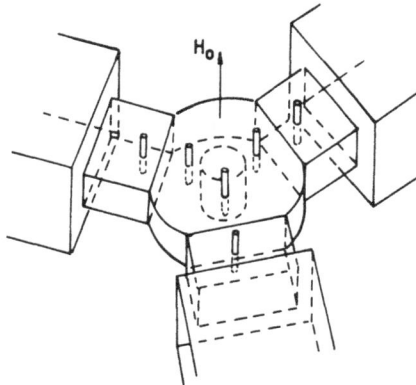

Fig. 6.13   Circulator with below-cutoff waveguide filters.

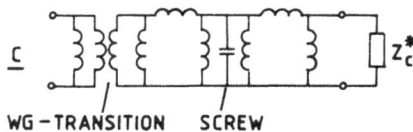

Fig. 6.14   Equivalent circuit.

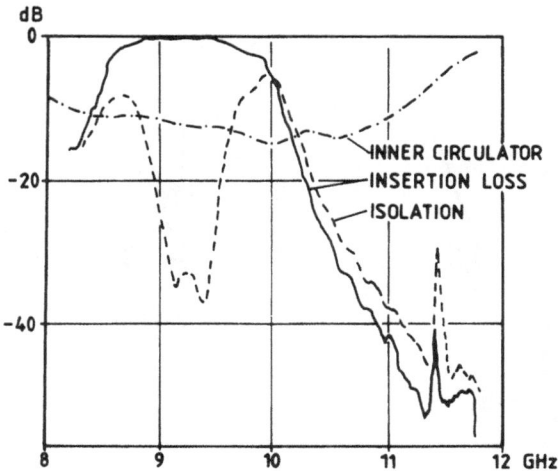

Fig. 6.15   Circulator performance.

GHz (Fig. 6.15). This restricted bandwidth (about 7%) was explained in terms of the limitations on bandwidth imposed by the junction with the propagating guide, which is a valid explanation if the junction is made into an additional resonator because its characteristics are imposed on a filter which is not designed to accommodate such a resonator. However, this is an unnecessary restriction as explained in Section 3.2.2 where it is shown that the resonator formed by tuning out the junction susceptance can be made a compatible part of the filter design rather than an additional unwanted resonator. The option also exists of avoiding the junction susceptance altogether by dielectric loading of the propagating section or by employing a ridge.

These conclusions led to the second configuration shown in Fig. 6.16 which was built in K-band guide. The operating frequency of the circulator was below the cutoff frequency of the K-band guide and also below that of the ferrite-loaded junction. Accordingly, the equivalent circuit of the ferrite-loaded resonator was deduced to involve capacitive elements (because the mode concerned in this case is TM_{01}) rather than inductive elements. The unit was tested by coupling it to a propagating X-band guide and impedance matching using simple quarter-wave transformers in the K-band guide (which was made to propagate by dielectric loading). The performance curves of the circulator are shown in Fig. 6.17 and demonstrate increased bandwidth — approximately double that of the earlier model. Not unreasonably Schieblich and Schünemann [18] claim that an increased bandwidth could be obtained with more sophisticated matching techniques.

From the foregoing work they conclude that two kinds of circulators are possible. In one the ferrite resonator is represented by a propagating mode in

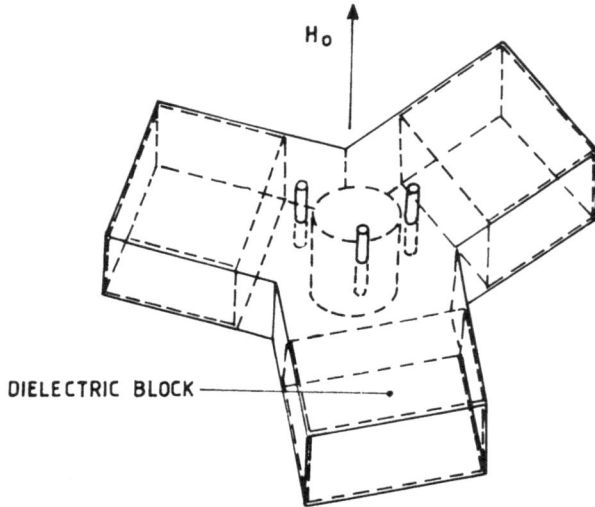

Fig. 6.16   Circulator using dielectric matching sections.

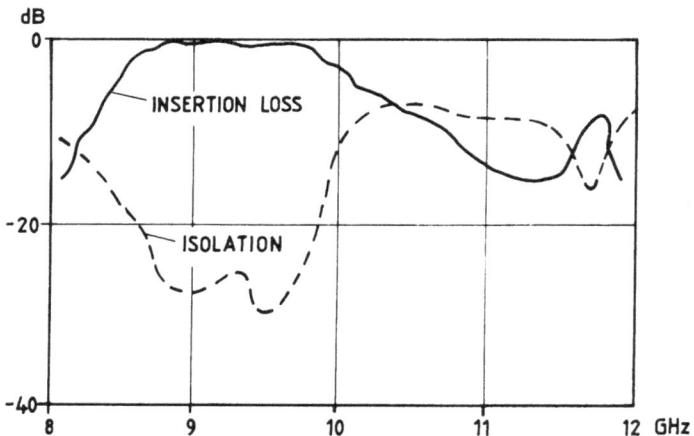

Fig. 6.17   Circulator performance.

the ferrite region but the output coupling circuits are in evanescent mode guide and this leads to a restricted bandwidth. From the above arguments this is correct but is the result of self-imposed limitations which are unnecessary. Their second conclusion that circulators exist in which the ferrite resonator is itself below cutoff and is brought to resonance by the adjacent coupling screws is original and warrants further investigation. Clearly, more research in this interesting field could produce further worthwhile advances.

References

1 Skedd, R. F. Waveguide junction circulators, *British Patent 1,190,547*, May 1970.
2 Skedd, R. F. and Craven, G. Magnetically tunable multisection bandpass filters in ferrite-loaded evanescent guide, *Electron. Lett.*, **3** (2), 62, February 1967.
3 Skedd, R. F. and Craven, G. New type of magnetically tunable multisection bandpass filter in ferrite-loaded evanescent waveguide, *IEEE Trans. Magn.*, **MAG-3** (3), September 1967.
4 Craven, G. F., Skedd, R. F. and Challans, J. C. Magnetically tunable bandpass filters in partially ferrite-loaded evanescent waveguide, *British Patent 1,340,162*, December 1973.
5 Craven, G. and Skedd, R. F. Magnetically controlled T–R switch, *British Patent 1,189,974*, April 1970.
6 Suhl, H. and Walker, L. B. Topics in guided wave propagation through gyromagnetic media, *Bell Syst. Tech. J., Monogr. 2322*, September 1954.
7 Lax, B. and Button, K. J. Theory of new ferrite modes in rectangular waveguide, *J. Appl. Phys.*, **26**, 1184, 1955.
8 Lax, B. and Button, K. J. New ferrite mode configurations and their applications, *J. Appl. Phys.*, **26**, 1186, 1955.
9 Craven, G. Waveguide bandpass filters using evanescent modes, *Electron. Lett.*, **2** (7), 251, July 1966.
10 Clarricoats, P. J. B. *Microwave Ferrites*, Chapter 5, Chapman and Hall, London, 1961.
11 Lax, B. and Button, K. J. *Microwave Ferrites and Ferrimagnetics*, McGraw-Hill, New York, 1962.
12 Craven, G. Waveguide below cutoff: a new type of microwave integrated circuit, *Microwave J.*, **13** (8), 51, August 1970.
13 Craven, G. Tuning techniques for multisection waveguide bandpass filters using evanescent modes, *Electron. Lett.*, **2** (11), 419, November 1966.
14 Craven, G. and Mok, C. K. The design of evanescent mode waveguide bandpass filters for a prescribed insertion loss, *IEEE Trans. Microwave Theory Tech.*, **MTT-19** (3), 295, March 1971.
15 Snyder, R. V. Stepped-ferrite tunable evanescent filters, *IEEE Trans. Microwave Theory Tech.*, **MTT-29** (4), 364, April 1981.
16 D'Ambrosio, A. Slimguide circulators in C, X, K and U bands, *Proc. Int. Microwave Symp., Ottawa, 1978*, p. 105.
17 D'Ambrosio, A. Realization of a K_u band uncooled parametric amplifier for spacecraft applications, *Proc. Microwave '73 Conf., Brighton, 1973*.
18 Schieblich, C. and Schünemann, K. Circulators with evanescent mode resonators, *Proc. 11th European Microwave Conf., Amsterdam, September 1981*, p. 394.

Chapter 7

Antenna Elements

7.1 Introduction

At a fairly early stage in this work the question arose of how comprehensive this alternative technique could be. For example, were antenna elements possible using cutoff guide resonators? Conceptually, it seemed that if the impedance of the tuned element was real when tuned to resonance and turned out to be a reasonable match to space, it was likely to radiate quite strongly. The image impedance approach employed in Chapter 1 shows quite clearly that the real impedance at band center is quite close to the numerical value of the wave impedance of the guide. Thus we would expect the guide as represented by a T-section to radiate, the main flaws in performance resulting from the incidental effects of discontinuity susceptance at the junction and from the mismatch between the real part of the guide impedance and that of space. Although these effects are not very large they militate against first-rate performance, but they can be removed by rearranging the resonator so that it ends in a capacitive element rather than an open-circuit section of guide. The resulting scheme [1, 2] shown in Fig. 7.1(a) then corresponds to a π-section resonator as defined in Chapter 1.

7.2 Design theory

In the equivalent circuit illustrated in Fig. 7.1(b) the coupling loop is represented by a transformer shunted by a tuning susceptance B_c. The admittance of the antenna aperture is denoted Y_a. A comparison of the structure in Fig. 7.1(b) with the corresponding impedance-matching structures discussed in Chapter 5 shows that the essential function of this network is to match the coaxial line impedance to that of free space. The main parameters in matching the impedance at the loop to that of space are the lengths of cutoff guide and dielectric plate in the aperture which match the real and imaginary parts respectively of the impedance. The permittivity and thickness of the dielectric section are the factors controlling the imaginary component.

121

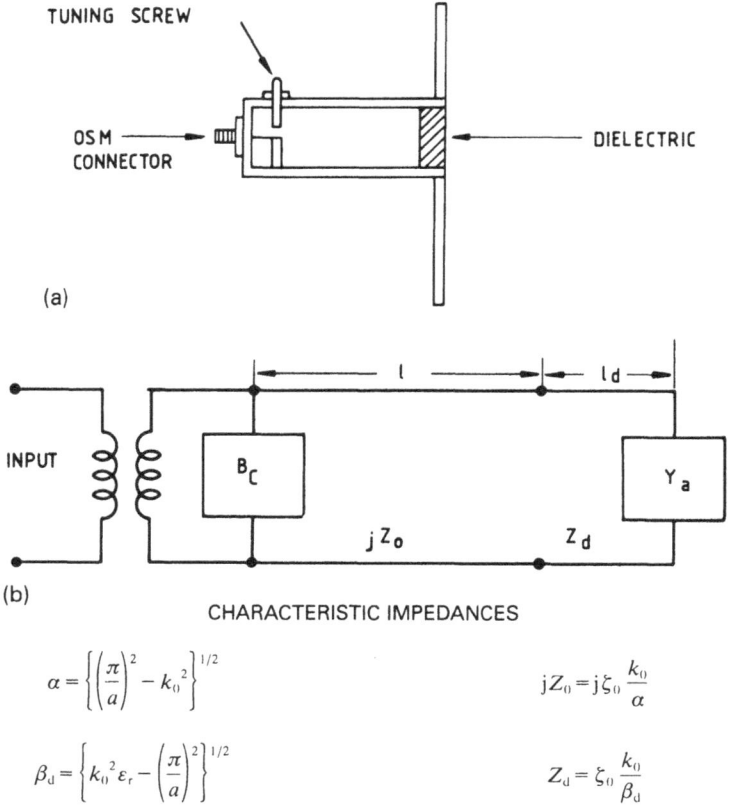

Fig. 7.1   (a) Antenna element (single section); (b) approximate equivalent circuit.

The aperture admittance Y_a, represents the first element which should be calculated. Lewin [3] has obtained an approximate expression for Y_a for a single waveguide, but the derivation is only accurate if both free space and the waveguide are filled with the same homogeneous isotropic dielectric. Since this condition is not fulfilled, a theoretical analysis [2] was performed which reduced the associated integrodifferential equation to a matrix equation by the method of moments [4]. The resulting equation was then solved on a computer and yielded accurate results for the terminal admittance. A feature of the analysis is that mutual coupling is automatically taken into account in multi-element arrays.

In common with all antennas the most important characteristic is the radiation pattern. Expressions for the two orthogonal components of the radiated electric field intensity in the far-field region are

$$E_\theta = A\, I(\theta, \phi) \cos \phi \tag{7.1}$$

$$E_\phi = - A \, I(\theta, \phi) \sin \phi \cos \theta \tag{7.2}$$

Here

$$A = j \left(\frac{k_0}{2\pi}\right)^2 \frac{2\pi \exp(-jk_0 r)}{k_0 r} \tag{7.3}$$

(r, θ, ϕ) are spherical coordinates, $k_0 = 2\pi/\lambda_0$,

$$I(\theta, \phi) = \iint\limits_{\text{aperture}} E_A(x, y) \exp(jwx + juy) \, dx \, dy \tag{7.4}$$

where

$$w = k_0 \sin \theta \cos \phi \tag{7.5}$$

$$u = k_0 \sin \theta \sin \phi \tag{7.6}$$

and $E_A(x, y)$ is the transverse electric field intensity at the aperture polarized in the direction of the x axis, which corresponds to $\phi = 0°$ and $\theta = 90°$.

For the present design, which consists of a single waveguide aperture of width a and height b it was established, subject to the restrictions $b < a$ and $a < \lambda_0/2$, that $I(\theta, \phi)$ was a slowly varying function over the whole hemispherical region corresponding to $\theta < \pm 90°$ and all values of ϕ.

The radiated power density function, by definition, is given by

$$P(r, \theta, \phi) = \frac{|E(r, \theta, \phi)|^2}{\zeta_0} \tag{7.7}$$

$$= \frac{|E_\theta(r, \theta, \phi)|^2 + |E_\phi(r, \theta, \phi)|^2}{\zeta_0} \tag{7.8}$$

where ζ_0 is the impedance of free space. Hence by substituting eqns (7.7) and (7.8) in (7.4) we obtain

$$P(r, \theta, \phi) = \frac{|A|^2 |I(\theta, \phi)|^2 (1 - \sin^2 \theta \sin^2 \phi)}{\zeta_0} \tag{7.9}$$

From (7.9) it can be seen that, subject to the above restrictions, the polar diagram is virtually independent of the aperture size and is approximately proportional to $1 - \sin^2 \theta \sin^2 \phi$. Thus the radiation in the E plane ($\phi = 0$) is effectively constant for all values of θ, whereas in the H plane ($\phi = \pi/2$) it varies at a rate which is at least as great as $\cos^2 \theta$.

A two-element array was considered as a method of reducing the very wide E-plane beam width. A pair of identical elements was constructed in WR75 waveguide for operation over the frequency band 5.0–5.25 GHz. The length of

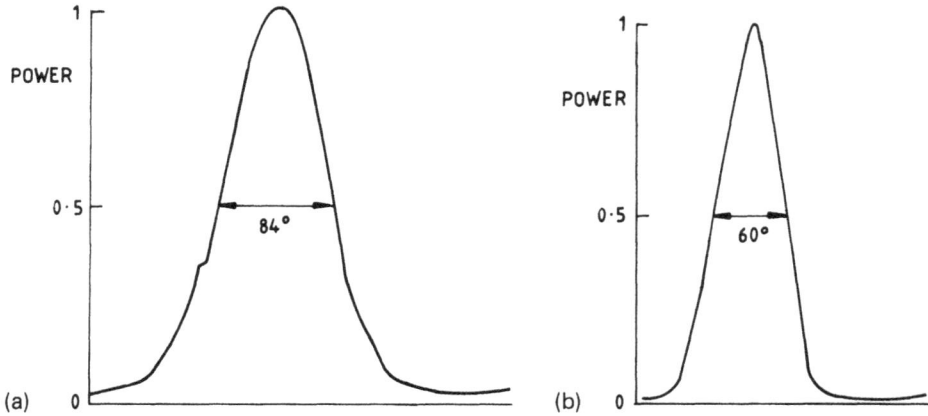

Fig. 7.2   (a) Measured *H*-plane polar diagram; (b) measured *E*-plane polar diagram.

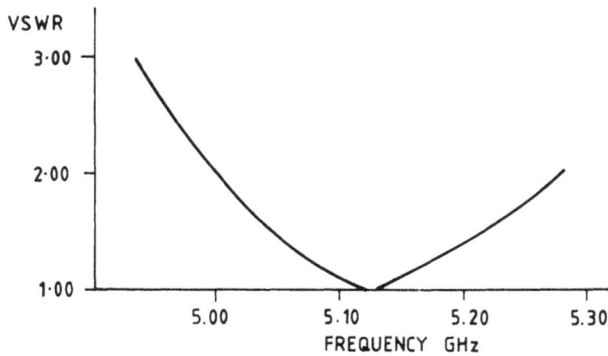

Fig. 7.3 Measured VSWR as a function of frequency.

each element was 2.54 cm. The measured polar diagrams shown in Figs 7.2(a) and 7.2(b) are in good agreement with theoretical expectations: the measured *E*- and *H*-plane beam widths were 60° and 84° respectively. The gain of the array was 8.7 dB and the midband insertion loss was 0.2 dB. The input VSWR was unity at the center frequency, increasing to a maximum of 1.9 at the band edges (Fig. 7.3). The curve is essentially that of a simple resonant circuit. More complex methods of matching which yield wider bandwidths are briefly described later.

The *H*-plane coupling between two differently sized units as a function of aperture size is shown in Fig. 7.4. The lower curve illustrates the coupling between WR75 guides for a spacing of 4.9 cm between centers, and the upper curve represents the corresponding case for WR187 guide with a spacing of

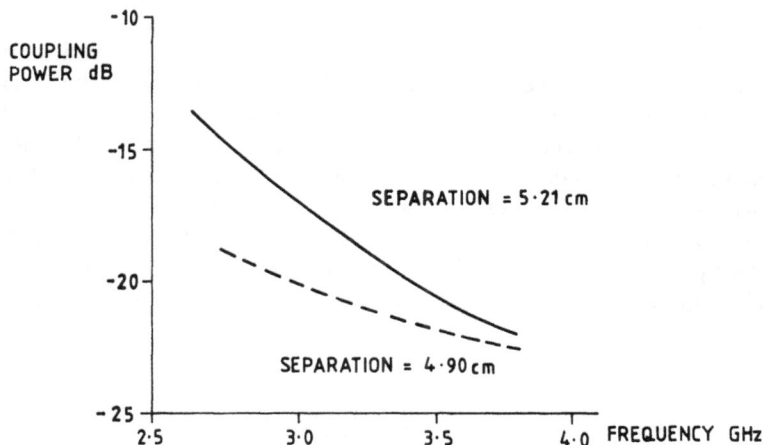

Fig. 7.4   Measured *H*-plane coupling.

5.21 cm. It can be seen that the average improvement for the smaller unit is 4.6 dB.

It is worth mentioning additional advantages that units of this type have in multi-element wide-angle scanning phased-array systems. One of the problems that occurs in antennas of this type, which use conventional above-cutoff guide elements, is that a "blind angle" can occur at large scan angles as a result of the interaction between the radiated beam and a higher evanescent mode. The phenomenon has been analyzed [5], and it has been shown that the effect is a result of evanescent mode resonance (known colloquially as the ghost mode) between the cutoff guide (at the higher-order mode) and the dielectric window. This is unlikely to occur in a system in which the guide is already below cutoff at the dominant mode because the resonance conditions are satisfied at that frequency by a dielectric which also serves as the aperture window. These conditions cannot easily be repeated at a higher-order mode because of the inherent mode selectivity of the resonance. In general, additional material with an exceptionally high dielectric constant would be required to make the higher-order mode resonance possible. Detailed investigation of this approach has been carried out and these advantages have been established.

7.3   Broader-band units

The system discussed in the previous section required only narrow-band elements, a specification which facilitated the size reduction achieved. As might be expected there is a trade-off between the size of the element and its bandwidth. This follows from the wave impedance of the cutoff guide which simply decreases as the waveguide is made smaller. Consequently, in addition to tuning it to make this impedance real, we shall have to transform the impedance over a greater range in order to match it to space. In this section we outline the design of a broader-band unit which was required for operation at L band.

The element was needed in a military aircraft landing system. The antenna, which was composed of these elements, was an array in which individual elements were switched sequentially so that the aircraft received a signal involving the Doppler effect from which the approach angle could be ascertained. An extremely wide-angle radiation pattern was needed. Switching was effected by

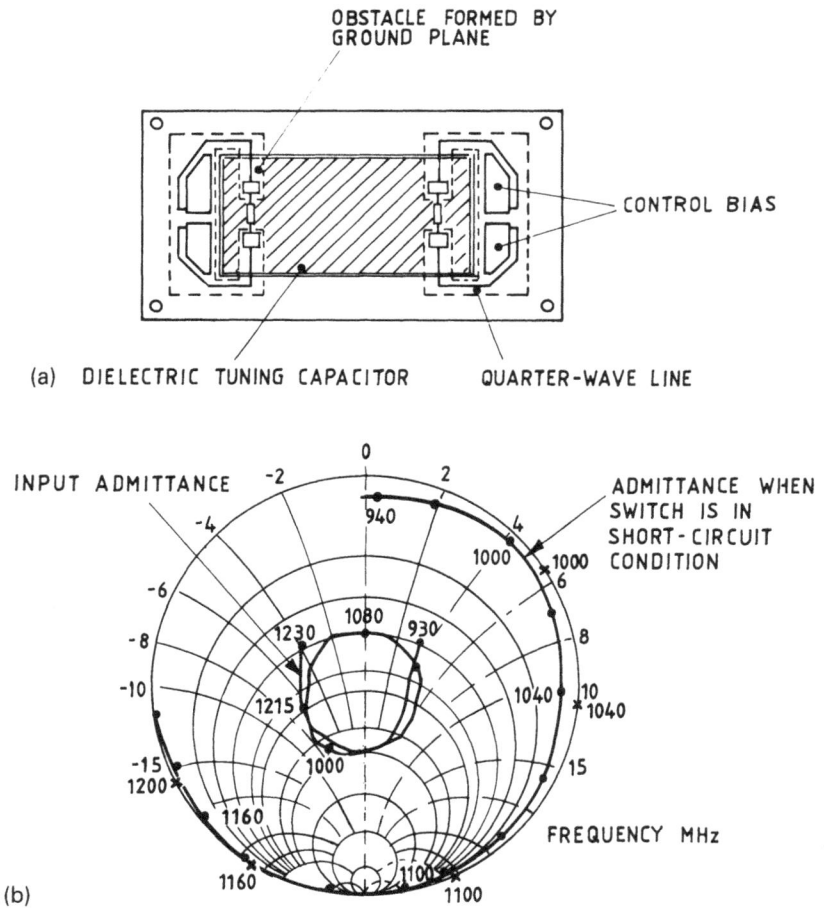

Fig. 7.5 Broader-band antenna: (a) aperture configuration; (b) measured admittance plot of the antenna.

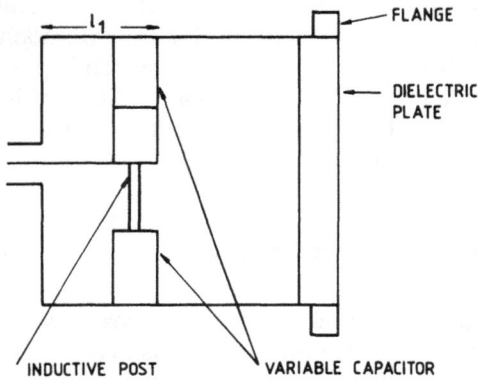

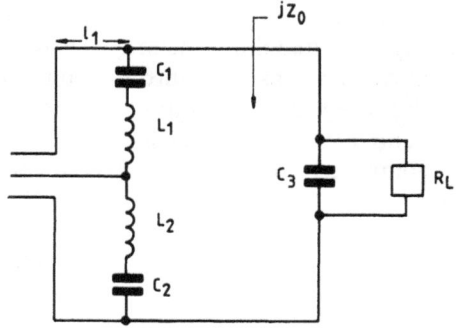

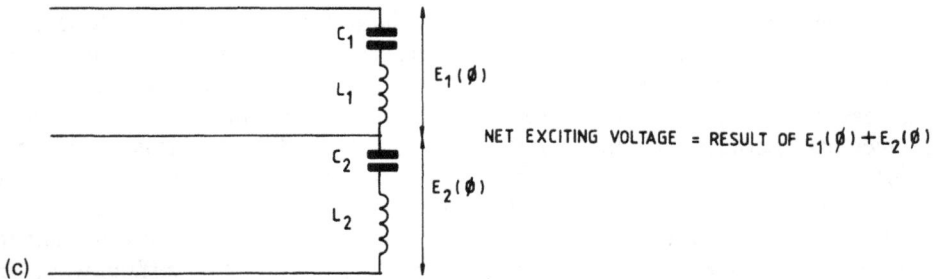

Fig. 7.6   Antenna input circuit: (a) construction; (b) equivalent circuit; (c) net exciting
voltages.

p–i–n diodes located at the coaxial input port, which left the element aperture
open; incidental coupling to other elements in the array then affected the radiation pattern. The purpose of the additional switch, located in the antenna
aperture and shown in Fig. 7.5(a), was to short-circuit the aperture whilst the
element was not excited and thus render the parasitic effects of the open
elements negligible during the cycle. Here we are not principally concerned with
the aperture switch, which was novel [6], but with the properties of the element.

The input match of the element is shown in Fig. 7.5(b). It was built in
WR510 guide ($f_c = 1.16$ GHz) and therefore cutoff was located in the frequency
range of operation. The modified input circuit is illustrated in Fig. 7.6 and was
developed empirically. The coupling is divided into two halves, each containing
an inductive element with the two inductances being of different magnitudes.
The net exciting field is the result of the difference in both magnitude and phase
between the currents flowing in each half of the circuit. From the equivalent
circuit we would expect two series resonances and a parallel resonance, result-

ing in an admittance curve which circles the chart as shown in Fig. 7.5(b). This complex matching technique was necessary to compensate for the additional frequency sensitivity introduced by the diode switch. Without the switch in position the total bandwidth was about 50% for the VSWR indicated. Thus wideband elements are possible which also yield a substantial reduction in size compared with a comparable element above cutoff.

7.4   Frequency-scan antenna

Frequency-scan phased-array antennas are directional antennas in which the angle of the radiated beam is a function of frequency. This is achieved by controlling the relative phase of excitation of individual elements in the array according to a predetermined phase–frequency characteristic. In order that the desired incremental phase change can be obtained with only a moderate frequency scan, it is necessary to employ a large total phase shift (a long electrical path) in each feeder section. Since the free-space distance between adjacent elements of the array is only of the order of half a wavelength, it is necessary to increase considerably the electrical length of the feeder sections which excite adjacent elements in the array. One way of doing this is to employ a tortuous path between the elements which is usually called a serpentine. A more sophisticated scheme which involves a slow-wave structure is to employ a surface wave. Existing serpentine structures are both complex and difficult to make, and similar objections exist to units which involve the surface wave. In Chapter 1 it is shown that the periodic structure which forms the basis of the work in this book is a slow-wave structure and should be ideal for this purpose since its construction is basically simple. The frequency-scan antenna [7] is shown in Fig. 7.7 and consists of a number of identical resonators with longitudinal slots cut in the broadside wall and located at a point midway between resonators. The relevant equation relating the phase can be expressed in the form [8]

$$\cosh \phi = 2 \left\{ \cosh^2\left(\frac{\alpha l}{2}\right) - Z_0 B_1 \sinh\left(\frac{\alpha l}{2}\right) \cosh\left(\frac{\alpha l}{2}\right) \right\} - 1 \qquad (7.10)$$

In terms of these concepts the network will transmit energy freely over the frequency range in which its image impedance is real. The limits of this range occur at

$$\cosh \phi = \pm 1 \qquad (7.11)$$

The importance of the relationship yielding the image transfer constant in eqn (7.10) is that it gives the phase at the planes a_1, b_1 a_2, b_2 etc. throughout the structure. It is worth noting that to obtain a corresponding phase variation in a propagating structure we would require a guide length of 20–100 times that of

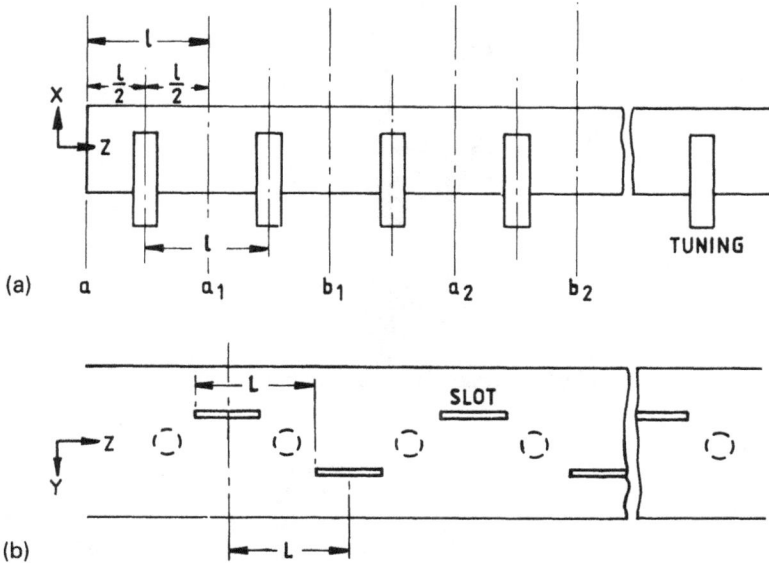

Fig. 7.7   Frequency-scan antenna: (a) elevation; (b) plan.

the corresponding slow-wave structure (depending on the bandwidth of the latter). The polar diagram of the antenna is therefore a function of the relative phase and amplitude of the fields in the slots. If ϕ is the phase shift per slot period L, the beam angle ψ_P with respect to the normal to the array (the x axis) is given by

$$\sin \psi_{P\pm} = \frac{\lambda_0}{2L} \left\{ \frac{\phi}{\pi} \pm (2P-1) \right\} \tag{7.12}$$

where $P = 1, 2, 3$ etc. A real beam exists in the direction ψ_P for each $|\sin \psi_P| \leqslant 1$. For

$$\frac{L}{\lambda_0} = \frac{1}{2^{1/2}} \tag{7.13}$$

the beam positions are those shown in Fig. 7.8.

The beam position for other angles ψ_P deserves consideration. For example, at the lower cutoff frequency there exist two beams ψ_{1+} and ψ_{1-}. As the frequency is increased to the upper cutoff frequency the beam ψ_{1-} moves to ψ° whilst ψ_{1+} moves to $+90^\circ$ and then into imaginary space. There exists a region of about 25° (the shaded area in Fig. 7.8) where ψ_{1-} is the only beam in space and this represents the useful range of angles for this slot period.

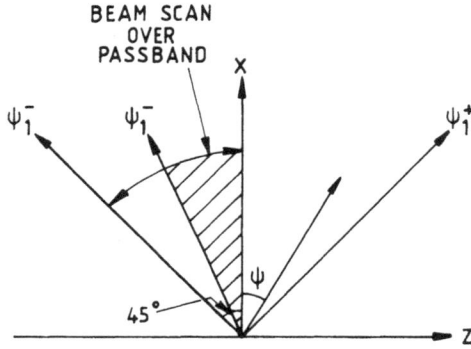

Fig. 7.8   Angular direction of the beam.

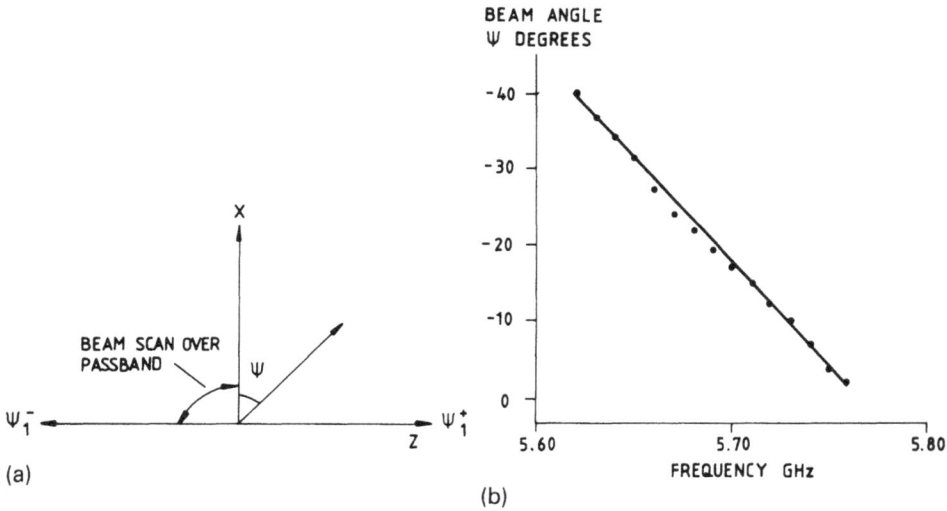

(a)

(b)

Fig. 7.9   (a) Range of the beam scan angle; (b) measured beam angle versus frequency.

Figure 7.9(a) shows the scan-angle range over the passband for the example

$$\frac{L}{\lambda_0} = \frac{1}{2} \tag{7.14}$$

At the lower cutoff frequency two beams exist in directions $\psi = \pm 90°$. As the frequency is increased ψ_{1+} moves beyond cutoff (into imaginary space) and ψ_{1-} increases, reaching $\psi°$ at the upper cutoff frequency. Thus, in this example ψ_{1-} is the only beam in real space. An antenna was built using 19 periods with

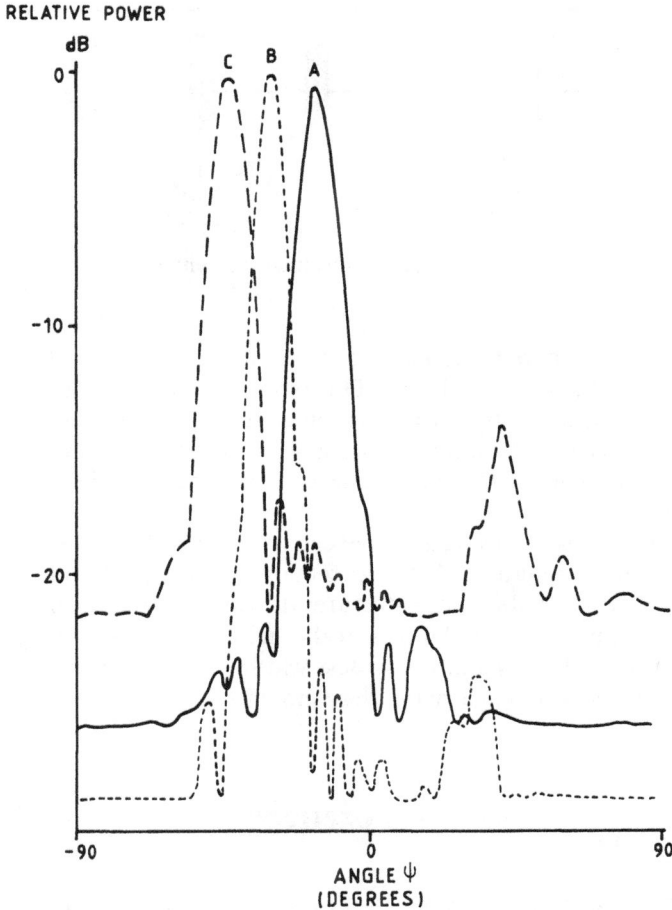

Fig. 7.10 Measured polar diagram.

$L/\lambda_0 = 1/2^{1/2}$; the measured beam angle is shown as a function of frequency in Fig. 7.9(b). This linear characteristic has a slope of 27.6° per 100 MHz at 5.7 GHz. The polar diagram is readily obtained once the field in the slots is known. For a structure with radiating slots located on the same side of the longitudinal guide center, the beam angles are given by

$$\sin \psi = \frac{\lambda_0}{2L}\left(\frac{\phi}{\pi} \pm 2P\right) \quad P = 0, 1, 2, \ldots \quad (7.15)$$

Typical measured polar diagrams showing the change in beam angle for frequency increments of 0.25 GHz from 5.25 to 5.75 GHz are shown in Fig.

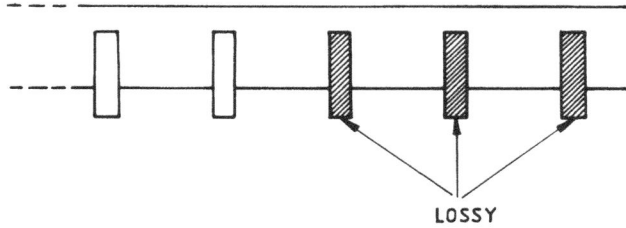

Fig. 7.11 Lossy terminating elements.

7.10, curves A, B and C. It can be seen that each polar diagram has an oppositely angled low power beam. This is the result of energy reflected from the end of the antenna. It can be prevented by using terminating resonators which include lossy screws; no radiating slots are included in the terminating end section. The end of the waveguide can then be closed by a short-circuit plate (Fig. 7.11).

The advantages of this frequency-scan antenna include the use of commercially available waveguide and the correction of waveguide tolerances by the tuning screws. It is simple in manufacture; the only operations involved are the drilling and tapping of screw holes and the milling of slots. Finally, it is lighter than competitive schemes and its modest width ($w < \lambda_0/2$) permits the stacking of linear arrays for wide-angle azimuth scanning.

References

1 Craven, G. F., Hockham, G. A. and Marsh, S. B. Waveguide antenna, *British Patent 1,312,506*, April 1973.
2 Hockham, G. A. Flush mounted aeronautical antenna, *Aerospace Antennas, IEEF Conf. Publ. (London)*, 77, 49, June 1971.
3 Lewin, L. *Advanced Theory of Waveguides*, Chapter 6, Iliffe, London, 1951.
4 Harrington, R. F. *Field Computation by Moment Methods*, Macmillan, New York, 1968.
5 Hansen, R. C. *Microwave Scanning Antennas*, Chapter 3, p. 316, Academic Press, New York, 1966.
6 Craven, G. and Thomas, R. R. Waveguide antenna switches using p-i-n diodes, *Electron. Lett.*, 13, 518, August 1977.
7 Hockham, G. A. and Craven, G. F. Waveguide antenna, *British Patent 1,409,749*, October 1975.
8 Craven, G. Waveguide below cutoff: a new type of microwave integrated circuit, *Microwave J.*, 51, August 1970.

Chapter 8

Complete Subsystems

8.1   Introduction

Among the electrical advantages of the technique, which have been outlined in earlier papers on the topic [1, 2], are the absence of the periodic resonances normally associated with distributed-constant networks, a Q factor which at its best falls only moderately short of that of the propagating waveguide and properties that, in general, facilitate broad-band performance. In addition, the practical merits of a cheap high precision medium for microwave components, which is inherently shielded and compact and readily permits the trading of loss for size, are advantages which have been endorsed by later workers in the field [3, 4]. These characteristics only apply individually to competing media to a more limited extent. It is also the experience of many engineers, once conceptual blockages have been overcome, that component development is much easier in waveguide below cutoff. The ease with which matching can be accomplished, as a result of the absence of the complex transformation that ordinarily occurs in transmission line problems and necessitates the use of the Smith chart, and the greater attenuation of (unwanted) higher-order evanescent modes in a medium already below cutoff at the dominant mode leads to less interaction between adjacent components.

The subsystems described in the following sections, with one exception, are working units of which we have at least some experience and therefore some insight. A number of subsystems developed by Schünemann et al. [4] are omitted merely because the descriptions in their paper are comparatively brief and therefore do not form a good basis for a review in the present text.

8.2   Frequency multipliers

Frequency multipliers in above-cutoff waveguide normally require a range of waveguide sizes in order to encompass the frequency range involved. This complicates the design by involving a number of junctions, and also introduces problems of parasitic passbands in the associated filters. These electrical difficulties are greatly mitigated by constructing the unit in a common waveguide size below cutoff [5]; this also has the merit of simplifying construction.

The diode mount employed is shown in Fig. 8.1, and the external extension to the crossbar is illustrated in Fig. 8.2. In this application the crossbar provides a useful impedance transformation, which is an obvious requirement in such a mount and facilitates matching at both input and output terminals of the varactor. An equivalent circuit for the crossbar, mount and diode in cutoff guide has been derived by Dahele and Hill [5] and is shown in Fig. 8.3. The variable capacitance and resistance of the diode are represented by C_c and R_c, and the length l depicts the combined inductance of the diode and its mount. The crossbar in cutoff guide takes on the properties of the inductances L_1 and L_2. The tuning screw in the guide is represented by C_a, and C_b accounts for stray capacitance effects which are associated with the crossbar. It will be noted that the impedance at the output terminals of the mount is much higher than that existing at the varactor terminals. This is a necessary condition because the complete circuit of Fig. 8.3 must resonate at both input and output frequencies, which the

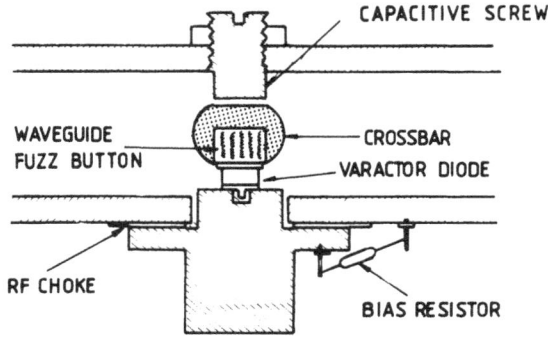

Fig. 8.1   Section through the diode mount.

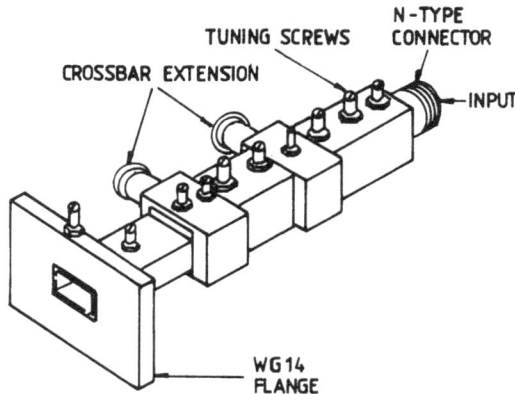

Fig. 8.2   1.9–7.6 GHz frequency multiplier consisting of two doublers in cascade.

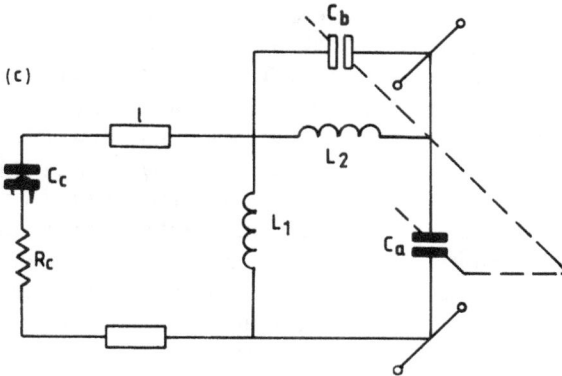

Fig. 8.3 Diode equivalent circuit.

low impedance of the varactor would otherwise make impossible. These resonances result from the out-of-band admittance of each filter in the transmission band of the other and have been discussed in detail earlier in the context of the conditions existing in diplexer design (Section 4.3).

The significance of the crossbar extension outside the guide as shown in Fig. 8.2 needs some explanation. When the waveguide is operated well below cutoff the crossbar effectively places a large inductive susceptance in shunt with the diode. The external extension then enables the admittance presented at the diode to be made zero, or any other value which facilitates the desired resonances, by an appropriate adjustment of the length. The crossbar also provides a useful conductive path to the waveguide walls for the heat generated in the diode.

The performance of two cascaded doublers yielding an output frequency in the 7.6 GHz band is shown in Fig. 8.4. The multiplier was designed as the

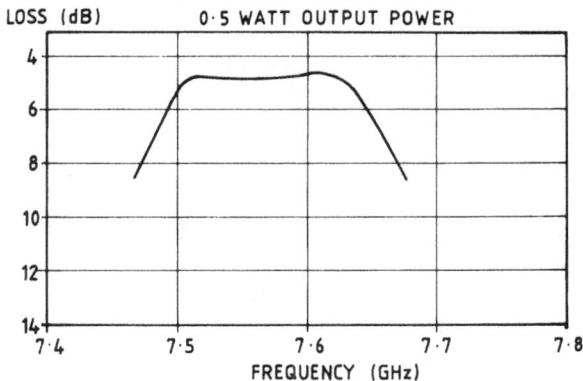

Fig. 8.4 Overall response of the 1.9–7.6 GHz frequency multiplier.

output stage of a 960-channel microwave link and carries the frequency-modulated output signal. Group-delay characteristics were within the required specifications as were other properties. Three models were built to cover the band 7.1–7.7 GHz. The estimated loss in the first diode and its mount was 1.5 dB, and the loss in the second diode–mount combination was 2.3 dB. The cross-bar has always been a difficult problem in analytical theory, but its convenience in the present application proved ample justification for the empirical approach that was necessary in design. The manufacture of the complete unit in a guide of common width resulted in a multiplier of compact construction; the unit shown was also free from parametric oscillations. Subsequently this unit was the basis for the manufacture of a number of multipliers used in NATO communications systems.

8.3   Microwave receiver [6]

Many of the units involved in a complete microwave receiver subsystem have been described in previous sections, but the combination of the individual components in a receiver is shown here (Fig. 8.5). It comprised a local oscillator filter, a mixer and a signal input filter. The signal input filter, which was designed for a bandwidth of about 0.5%, also included a harmonic suppression filter. This was necessary because the signal filter was transparent at the sum frequency generated by the mixer diode. In the absence of the additional filter, transmission of this frequency component through the main filter and its subsequent reflection back to the mixer ruined the group-delay characteristic. The importance of loss in this filter was considerable because of its effect on the

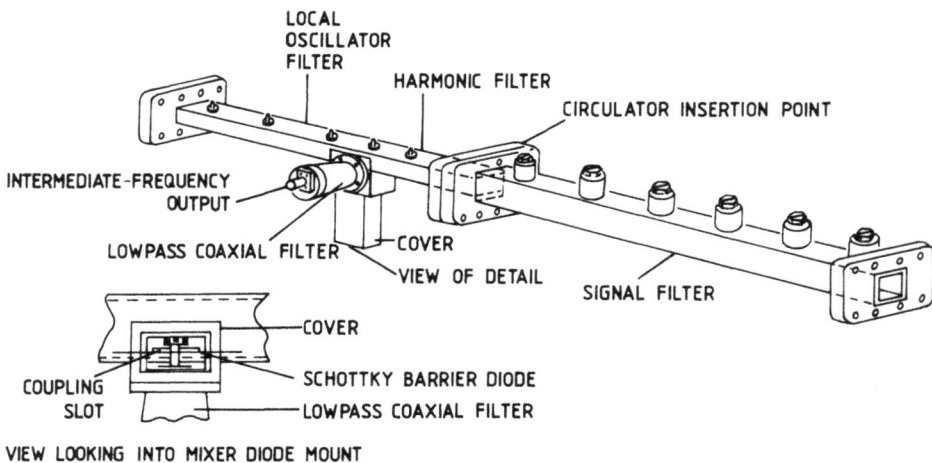

Fig. 8.5   Construction of a 7 GHz down-converter using cutoff waveguide.

overall noise figure, and a quartz-rod-tuned filter was employed. Inductively loaded versions were developed for applications where the unloaded model would be too long. The use of image-recovery techniques, including short-circuit termination of the sum frequency, resulted in a better performance than the conventional waveguide specification. The improvement of 0.7 dB was nevertheless considerably less than should have been possible, and after extensive checks on resonator Q had shown that this was satisfactory the problem was considered to be located in the lowpass filter and the intermediate-frequency matching circuit. The local oscillator filter was a simple three-resonator unit employing capacitive screws. In some field applications mixer diode replacement without the necessary facilities for rematching the diode was considered a possibility. To provide for this contingency an isolator was introduced at the flange point shown in the diagram. In addition, a simple local oscillator rejection filter was interposed between the isolator and the mixer in order to avoid wastage of local oscillator power in the isolator.

8.4 Marecs diplexer [7]

The Marecs diplexer provides the necessary multiplexing of transmitter and receiver onto a common antenna. It existed in the satellite as a subsystem in its own right and, although a purely passive network, is thought worthy of inclusion in this chapter because it illustrates the application of the technique to an unusually difficult specification. It is also of interest because the unit was part of a commercially successful satellite in the Marisat system. The difficult specifications arose directly from the ambitious nature of the satellite: its need to provide communication with ships in L band which could only employ dishes of limited size, the decision to develop a high power transistor amplifier and the usual restrictions on mass which tend to influence the design of all satellite components. These problems have been discussed in greater detail elsewhere [8].

The power involved and the operating frequency introduced the problem of multipaction. Briefly, multipaction is a secondary emission phenomenon which occurs in a hard vacuum at moderate to high power levels. It has been studied in detail in a number of papers [9, 10]. Optimum conditions for exciting multipaction require an electron motion which is in phase with the associated rf field. An electron traveling across a gap between two walls releases secondary electrons on impact with the wall. If this occurs at the moment when the exciting field is zero, these electrons are effectively primary electrons on the next half-cycle and release further electrons on collision with the opposite wall. This results in a build-up of current between the two surfaces which is known as multipactor breakdown. Clearly, the distance between the electrodes is critical, the shape of the electrodes is of some importance and the applied voltage — the sum total of any dc voltage and rf voltage — is important. A sufficiently high dc bias will frequently suppress multipaction, although its use in many applications can be very inconvenient. Multipaction is also known to occur in valves from time to time depending on the combination of voltages involved.

This brief digression into multipaction is considered desirable because of the overriding importance that this phenomenon had on the design of the diplexer. The nominal power output of the transistor power amplifier was 70 W at a frequency of 1540 MHz, a band in which multipaction is common. It was necessary to allow for the power peaks that occur during multicarrier operation and this raised the peak power to 400 W. Large safety factors were introduced because of the ultimately destructive nature of multipaction, and this resulted in a final test power of approximately 2.5 kW. Other factors of importance were the low loss required in the transmit path (loss less than 0.275 dB), the attenuation of more than 110 dB in the receive filter and the high rejection of the harmonic frequencies of the transmitter. The overall system trade-offs placed a strong emphasis on low loss, but within that specification the requirement was for moderate dimensions and low mass (nominally not in excess of 2 kg).

Initially, quartz dielectric blocks were considered because of their low loss, but they were finally rejected because of their high mass and the extensive testing for multipaction that would have been necessary. Capacitive rods were finally selected as the most practical solution, but because of multipaction it was essential to construct the transmit filter in square guide. The receive filter presented fewer problems and was built in rectangular guide. Because of the need for a relatively low mass, aluminum was selected as the manufacturing material, but it was recognized that this material would have to be silver plated in order to achieve the lowest possible loss. An additional factor in reducing the mass was inductive loading, which is described in an earlier chapter, in order to shorten both filters. Chemical milling, i.e. reducing the thickness of the wave-guide wall in selected areas, was employed as a means of reducing the mass still further.

The cutoff frequency for the guide was chosen to be 2.08 GHz. An additional feature was harmonic suppression in the transmitter output circuit. Although this rejection could be comprehensively obtained with a lowpass filter, the loss intrinsic in the more general approach to the problem would have been greater and there would have been the ever-present possibility of multipaction in the unit. For this reason, spot rejection of the second harmonic (two sections) and third harmonic (one section) — the only frequencies considered to be important — was obtained more economically by locating loosely coupled rejection circuits in the input connector of the transmit filter.

The overall performance is shown in Fig. 8.6. All the specifications were met, including the loss in the transmit path (loss less than 0.25 dB). The silver plating achieved here was of very high quality. Power handling up to 2.5 kW without breakdown of any kind was attained. The reduction in wall thickness by chemical milling is clearly evident in Fig. 8.7, as is the physical strengthening (essential to withstand the vibration tests) in the input connector to the square transmit filter. Additional strengthening built into the flange joining the transmit

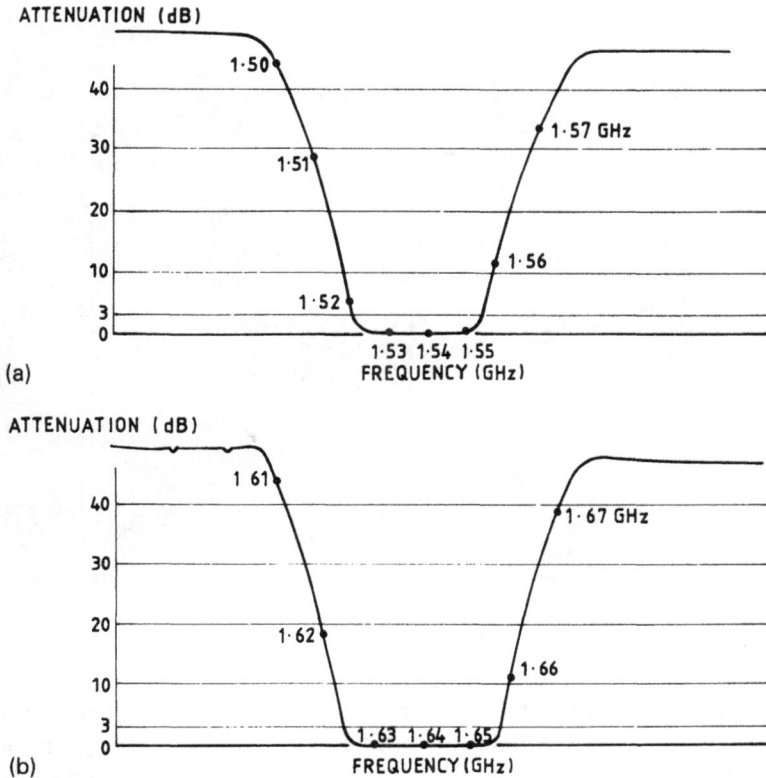

Fig. 8.6 Performance of a Marecs diplexer: (a) transmit path; (b) receive path.

and receive filters can also be seen. The overall length and width of the diplexer were 96 cm and 7.62 cm respectively. The guide height of the five-resonator transmit filter was 7.62 cm, and the corresponding dimension of the six-resonator receive filter was 3.87 cm. The overall mass of the unit was 1.6 kg.

8.5 Microwave link up-converter [11]

The absence of periodic resonances can be very useful in parametric up-converters. In principle an up-converter of this type requires two filters, an appropriate varactor diode, a filter for the signal frequency and a pump source with significant power (in excess of 1 W). Units of this type have been built in propagating guide but in microwave links of high channel capacity the bandwidth becomes quite broad at the signal frequency (intermediate frequency),

Fig. 8.7   A Marecs diplexer.

commonly as wide as 50% at the 3 dB points. Thus the signal circuit can support harmonics of the signal frequency, resulting in a spectrum of frequencies at the varactor diode. In propagating guide the presence of the waveguide obstacles (which are part of the filters) on each side of the diode provides the opportunity for additional resonances of high Q factor in this band of frequencies, and absorption effects occur in the output response at these resonances. This configuration has therefore lost popularity in such applications and has been replaced by the reflection type of up-converter which avoids these problems by resistively terminating unwanted modulation products. However, the dissipative termination of these products represents a loss of efficiency.

Reversion to the two-filter configuration is possible using the scheme illustrated in Fig. 8.8. In the diagram the pump frequency is introduced via the propagating guide on the right which includes a three-resonator filter. The six-resonator output filter combines with the pump frequency filter to produce wholly reactive terminations of a monotonic type. Additional resonances cannot occur in such a circuit. Practical details of the diode mounts are somewhat similar to those of the multiplier described in Section 8.2, with a crossbar mount being employed.

The performance is illustrated in Fig. 8.9. The power output was 1.1 W, and the total loss between pump and output port was 2.6 dB which included loss in the filters. In addition to having good performance the unit is compact and is simple and cheap to construct.

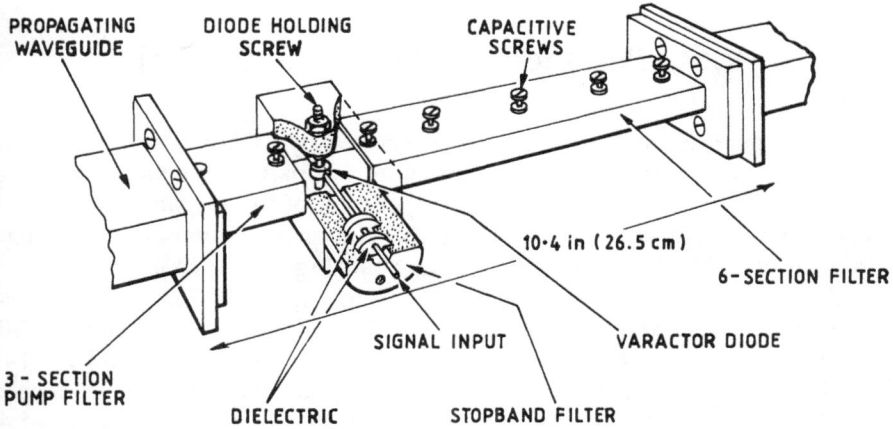

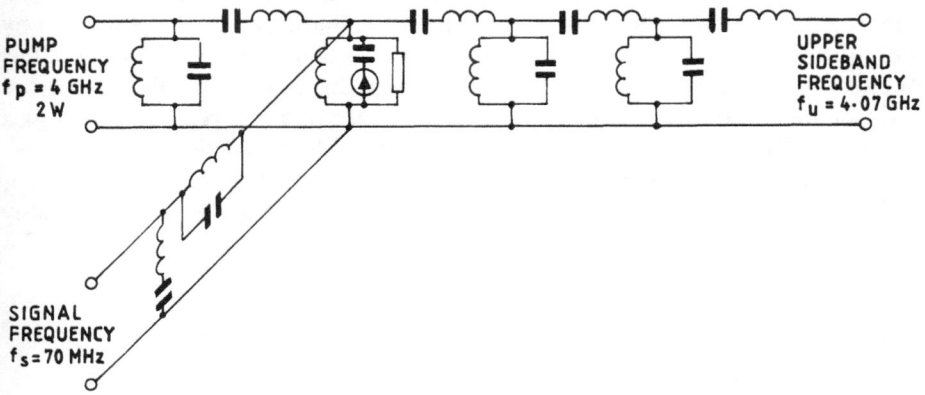

Fig. 8.8 Up-converter: (a) construction; (b) equivalent circuit.

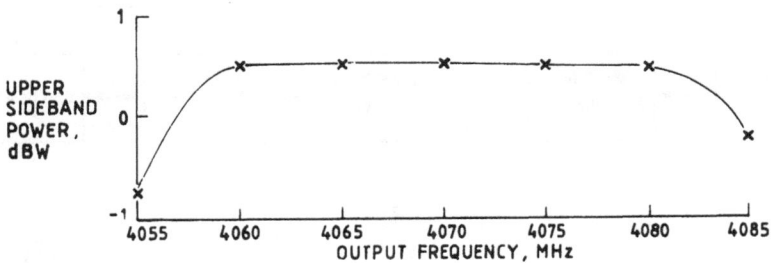

Fig. 8.9 Up-converter performance.

8.6   Parametric amplifier [3, 12, 13]

A parametric amplifier has been described by D'Ambrosio [3] and was developed by GTE, Milan, Italy. The amplifier was part of an experimental high powered satellite developed with the cooperation of the Department of Communications of Canada and involving the National Aeronautics and Space Administration (NASA) and the European Space Agency (ESA).

A block diagram of the unit is shown in Fig. 8.10. The amplifier was uncooled, but was required to yield a gain of 13 dB over a bandwidth of 300 MHz with a noise figure not exceeding 3.3 dB in the frequency band 14–14.3 GHz. Restrictions on mass and power consumption limited these to, respectively, less than 450 g and 3 W from a 15 V supply. A pump frequency of 40 GHz was required and preliminary consideration was given to a Gunn oscillator at this frequency and then to a Gunn oscillator located at 10 GHz followed by a quadrupler. The final choice was a transistor oscillator at 2.5 GHz followed by multiplication by a factor of 16. The microwave components of the amplifier were completely realized in below-cutoff waveguide, as also were the circulators and isolators. D'Ambrosio quotes the performance of a five-port circulator as a loss of 0.2 dB (between input and paramp ports) with an isolation of 50 dB over a band exceeding 500 MHz. The mass was 70 g and the performance held over a temperature range of − 40 to 70 °C. The use of cutoff guide led to an assembly

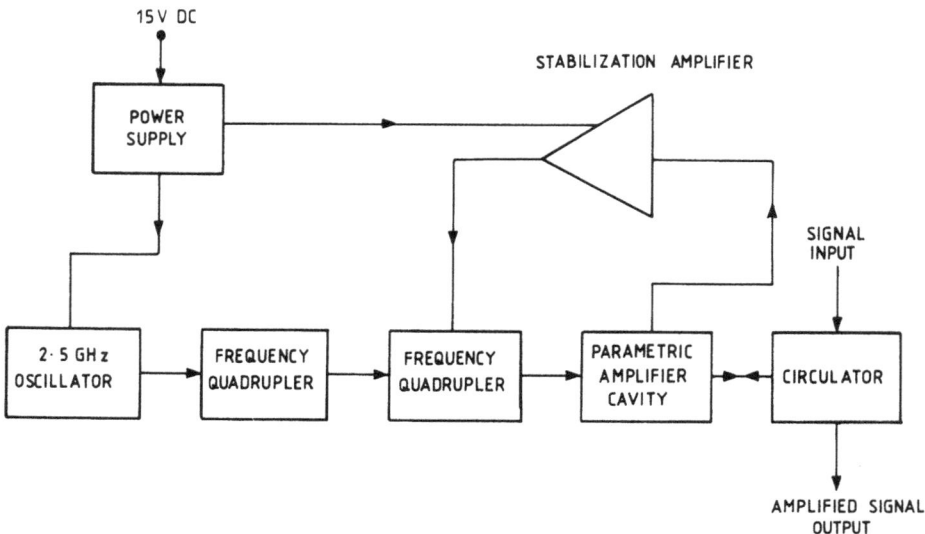

Fig. 8.10   14 GHz parametric amplifier.

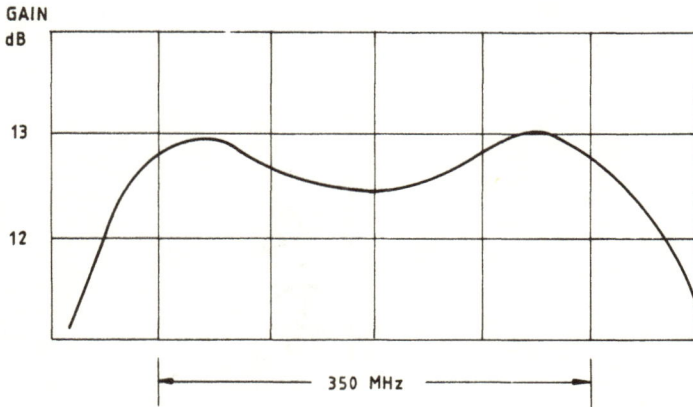

Fig. 8.11 Performance of the parametric amplifier.

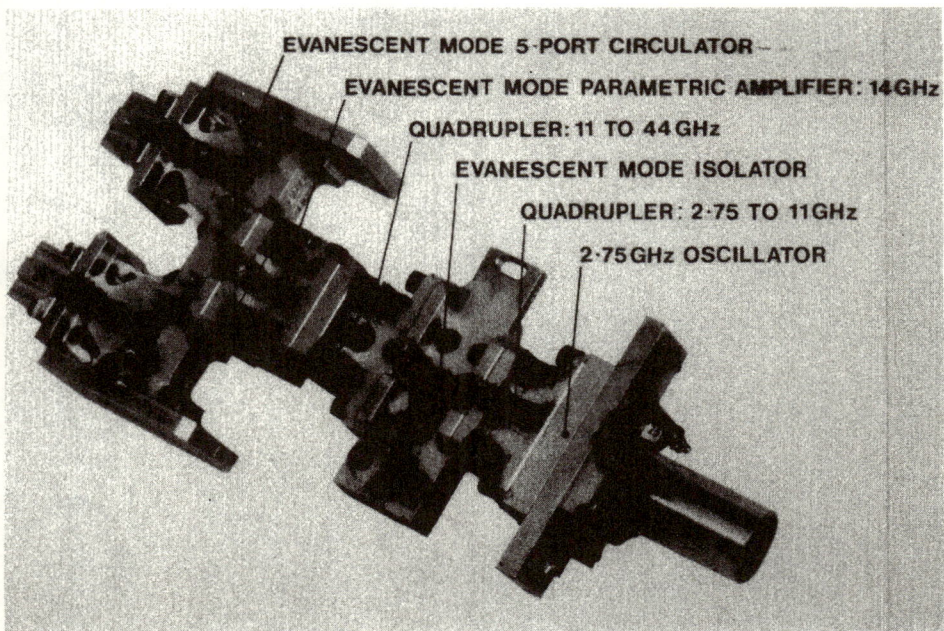

Fig. 8.12 The parametric amplifier.

which was lighter and more robust than a corresponding subsystem employing thin-film technology.

A typical performance curve (Fig. 8.11) illustrates the gain of the amplifier over the design bandwidth. With its use of multipliers, wideband circulators and

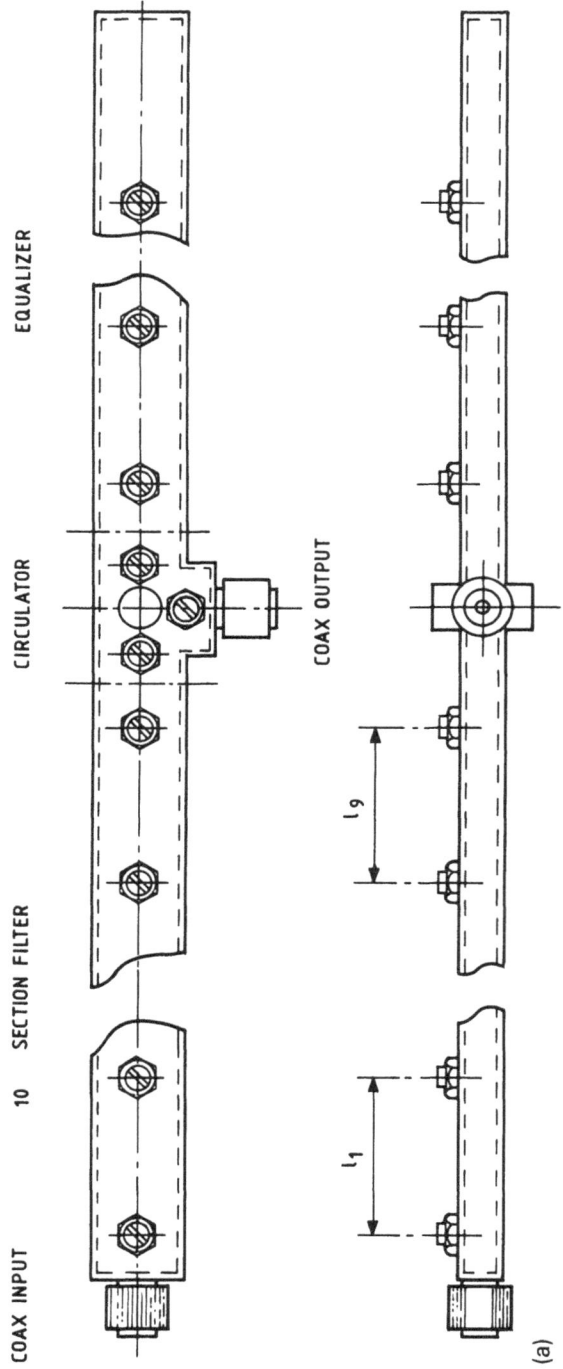

(a)

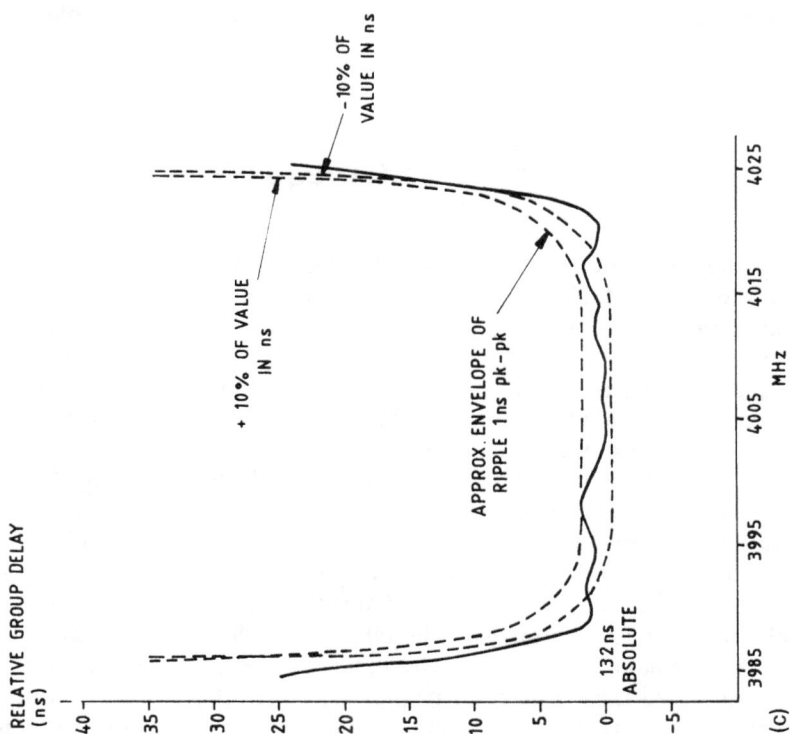

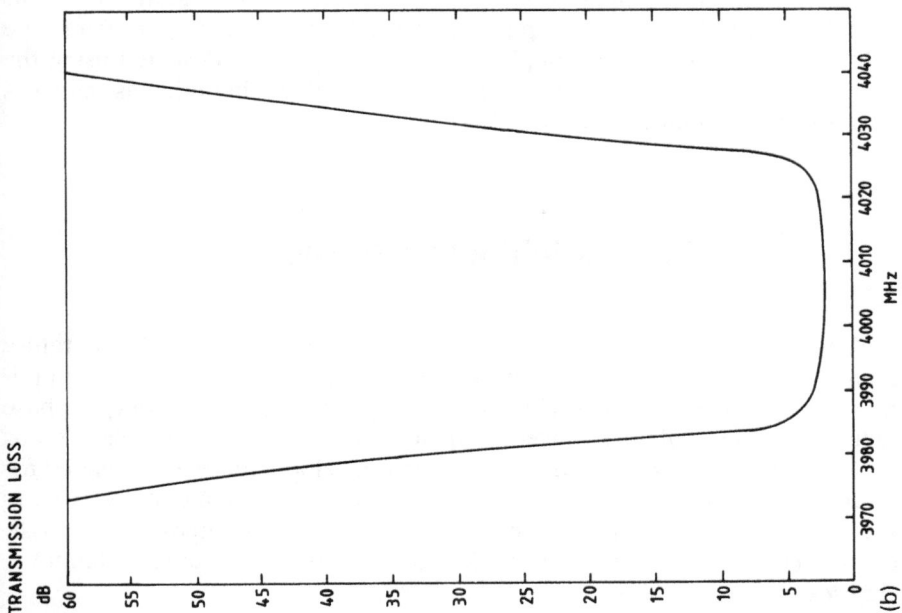

Fig. 8.13 (a) Filter–equalizer using a circulator; (b) the transmission response of the filter–equalizer; (c) the group delay of the filter–equalizer.

oscillators, the amplifier represents one of the outstanding applications of the technique to microwave subsystems (Fig. 8.12).

8.7   Filter–equalizer

The filter–equalizer was developed fairly early in the program (1968) and was intended to highlight the advantages of the technique where spectacular reductions in mass are necessary. The unit is shown in Fig. 8.13 and illustrates the basic scheme — a 10-resonator filter (0.01 dB Chebyshev) followed by a circulator which connects to a three-resonator equalizer. The filter–equalizer has been described in greater detail in previous papers [14, 15] and only its chief characteristics are summarized here.

The insertion loss of the complete unit is illustrated in Fig. 8.13(b) and the group-delay characteristic is shown in Fig. 8.13(c). The latter property was well within the required specifications, but the insertion loss was slightly in excess of the specified 4 dB. More seriously, the gain slope (a measure of the flatness of the passband) exceeded the required amount. Both these deficiencies were due in the main to shortcomings in the circulator, which was an early design. The latter performance of five-port circulators, as described in Section 8.6, would have remedied these passband ripples. The reduction in mass compared with the propagating-guide version was spectacular: a mass of only 340 g for a complete filter–equalizer compared with 4000 g for the conventional unit. This extremely large reduction was in part facilitated by a different approach to the equalizer. A three-resonator (coupled-resonator) equalizer, designed using the principles outlined by Merlo [16], has inherent advantages in both mass and loss over cascaded constant-resistance units [17].

8.8   Additional subsystems

A variety of additional subsystems have been described in the literature and are briefly mentioned. They include the following: injection-locked amplifiers [4], mechanically tuned oscillators [4], cavity-stabilized oscillators [4], phase modulators [4] and balanced mixers [18]. Some of the work on oscillators was preceded by that of Conlon et al. [19], but their developments were intended for different applications. In addition much unpublished work, some of it very early in the development of evanescent mode technology, was carried out on classified projects by Ketteringham at Marconi Defence Systems, Stanmore, Middlesex.

References

1 Craven, G., Mok, C. K. and Skedd, R. F. Integrated microwave systems using evanescent mode waveguide components, *Proc. European Microwave Conf., IEE Conf. Publ.*, **58**, 285, September 1969.

2 Craven, G. Waveguide below cutoff: a new type of microwave integrated circuit, *Microwave J.*, **13** (8), 51, August 1970.

3 D'Ambrosio, A. Realization of a Ku band uncooled parametric amplifier for spacecraft applications, *Proc. Microwave '73 Conf., Brighton, 1973*, p. 495.

4 Schünemann, K., Knockel, R. and Begemann, G. Components for MIC with evanescent mode resonators, *IEEE Trans. Microwave Theory Tech.*, **MTT-25**, 1026, 1977.

5 Dahele, J. and Hill, D. R. Slimguide frequency multipliers for microwave applications, *Electr. Commun.*, **49** (1), 65, 1974.

6 Craven, G. Slimguide microwave components. *Electr. Commun.*, **47** (4), 254, 1972.

7 Craven, G. and Radcliffe, C. Microwave filters for communication satellites, *IEE Microwaves Opt. Acoust.*, **2** (5), 167.

8 Robins, W.P. Communication via geostationary satellites, *GEC J. Sci. Technol.*, **40** (1), 2, 1973.

9 Hughes Aircraft Co. The study of multipactor breakdown in space electronic systems, *NASA Contract. Rep. CR-448*, July 1966.

10 Clancy, P. Multipactor control in microwave space systems, *Microwave J.*, **21**, 77, March 1978.

11 Kwiatkowski, W., Arthanayake, T. and Knight, V. H. Efficient high-level upconverter for radio link, *Electron. Lett.*, **6** (19), 625, September 1970.

12 D'Ambrosio, A. A SHF parametric amplifier for space applications, *Alta Freq.*, **43** (10), 883, 1974.

13 D'Ambrosio, A. Design and realization of a parametric amplifier for satellite communication earth station, *Alta Freq.*, **40** (6), 534, June 1971.

14 Craven, G. and Mok, C. K. Lightweight microwave components for communication satellites, *Symp. on Long Term Prospects for Satellite Communication, Istituto Internazionale della Communicazioni, Genoa, 1971*.

15 Craven, G. Slimguide microwave components, *Electr. Commun.*, **47** (4), 245, 1972.

16 Merlo, D. Development of group-delay equalizers at 4 GHz, *Proc. Inst. Electr. Eng.*, **112** (2), 289, February 1967.

17 Chen, W. H. *Linear Network Design and Synthesis*, Section 14.6, McGraw-Hill, New York, 1964.

18 Begemann, G. and Peters, C. X-band balanced mixer using evanescent mode circuitry, *Electron. Lett.*, **14** (24), 748, 23 November 1978.

19 Conlon, R. F. B., Cooke, R. E. and Heeks, J. S. Broadband injection locking of Gunn oscillators, *Proc. 3rd European Microwave Conf., Brussels, 1973*, Session A.3.1.

Chapter 9

Measurements

9.1   Introduction

Microwave impedance measurement usually involves observing the ratio of the reflected waves to incident waves when an admittance or impedance is introduced into an otherwise matched transmission line circuit. In its simplest form this necessitates measuring the standing wave pattern that results using the familiar slotted line; more often nowadays the reflected wave is measured using a directional coupler which is part of a network analyzer. The admittance of the obstacle can then be deduced by appropriate methods involving the Smith chart.

Free propagation does not occur below cutoff and therefore the essential precondition for admittance measurement does not exist. Thus, in the absence of standing wave patterns, what can be done to determine the admittance of an obstacle which has been introduced into the guide? Some of the methods which have been employed are described in this chapter.

9.2   Mok's method [1]

A method was devised by Mok comparatively early in the investigations when a measurement technique was essential in order to check the behavior of obstacles below cutoff. More convenient and accurate techniques have been developed since, but in our opinion the advantage that Mok's method is equally applicable both above and below cutoff is almost unique. Thus a relationship can be established between measurements in the two regions, which yields a physical link which would not otherwise exist. This was particularly important in the early phases of the work. To consider an example the susceptance of an asymmetrical capacitive obstacles in a waveguide is (Section 2.2)

$$B \approx \frac{2b\omega\varepsilon_0}{\pi}\left[\left\{\varepsilon_r - \left(\frac{\pi}{ak}\right)^2\right\} + \left\{1 - \left(\frac{\pi}{ak}\right)^2\right\}\right]\ln\left\{\mathrm{cosec}\left(\frac{\pi d}{2b}\right)\right\} \qquad (9.1)$$

The corresponding susceptance B_f and B_u in dielectric-filled guide and unfilled guide respectively are

$$B_f = \frac{4bw\varepsilon_0}{\pi}\left\{\varepsilon_r - \left(\frac{\pi}{ak}\right)^2\right\} \ln\left\{\text{cosec}\left(\frac{\pi d}{2b}\right)\right\}$$ (9.1a)

$$B_u = \frac{4bw\varepsilon_0}{\pi}\left\{1 - \left(\frac{\pi}{ak}\right)^2\right\} \ln\left\{\text{cosec}\left(\frac{\pi d}{2b}\right)\right\}$$ (9.1b)

Consequently

$$B = \frac{B_f}{2} + \frac{B_u}{2}$$ (9.2)

These equations can be used as the basis of impedance measurements using a slotted line or one of the more sophisticated techniques. The experimental essentials are outlined in Figs 9.1(b) and 9.1(c). The filled guide supports propagation, and two measurements are made, one with a short-circuit located a suitable distance from the obstacle and the other with a long section of cutoff guide behind the obstacle. We then have the following equations:

$$Y_{i1} = jB_f + Y_{sc}$$ (9.3)

$$Y_{i2} = \frac{j}{2}(B_f + B_u) + Y_e$$ (9.4)

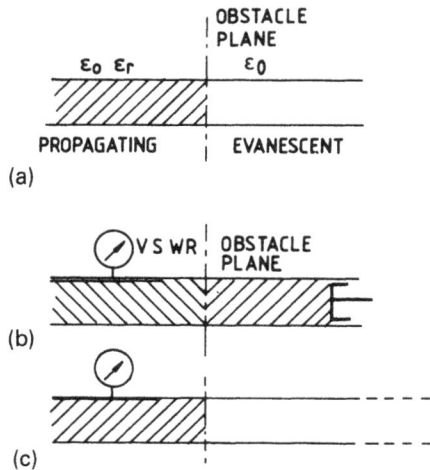

Fig. 9.1 Mok's method of susceptance measurement.

which yield

$$jB_u = 2(Y_{i2} - Y_e) - (Y_{i1} - Y_{sc}) \qquad (9.5)$$

where Y_{sc} is the admittance of the short-circuited propagating (filled) guide, Y_e is the characteristic admittance of the evanescent (unfilled) guide, jB_f is the admittance of the obstacle in the filled guide and jB_u is the admittance of the obstacle in the unfilled guide.

The advantage mentioned earlier is illustrated in the curves shown in Sections 2.3 and 2.5. Here, measurements taken adjacent to cutoff establish the continuity of the susceptance–frequency relation when the susceptance is normalized with respect to the admittance of free space instead of the more common guide admittance. A somewhat similar approach by Schieblich and Schünemann, which has the advantage of measuring the ratio of the fields at the input and output of the guide section, is worthy of study [2].

9.3 Insertion loss

One of the methods commonly used above cutoff to measure the susceptance of an obstacle in shunt with the guide is to determine the insertion loss when the obstacle is interposed between the source and a matched load. A similar technique can be employed below cutoff, but the insertion loss can be positive or negative depending on whether the obstacle is inductive or capacitive. The method is illustrated in Fig. 9.2.

The first measurement gives the attenuation of the guide in the absence of the obstacle and the second yields the corresponding attenuation with the obstacle in position. A straightforward way of determining the obstacle susceptance B is simply from curves obtained from a computer evaluation of the matrix product

$$
\begin{vmatrix} 1 & R \\ 0 & 1 \end{vmatrix}
\begin{vmatrix} \cosh(\alpha l) & jZ_0 \sinh(\alpha l) \\ \dfrac{\sinh(\alpha l)}{jZ_0} & \cosh(\alpha l) \end{vmatrix}
$$

$$
\times \begin{vmatrix} 1 & 0 \\ jB & 1 \end{vmatrix}
\begin{vmatrix} \cosh(\alpha l) & jZ_0 \sin(\alpha l) \\ \dfrac{\sinh(\alpha l)}{jZ_0} & \cosh(\alpha l) \end{vmatrix}
\begin{vmatrix} 1 & 0 \\ \dfrac{1}{R} & 1 \end{vmatrix}
\qquad (9.6)
$$

which gives the attenuation of the guide as a function of susceptance. The foregoing matrix product illustrates a capacitive obstacle. Clearly, when susceptances that have values which will yield near-resonant conditions are to be

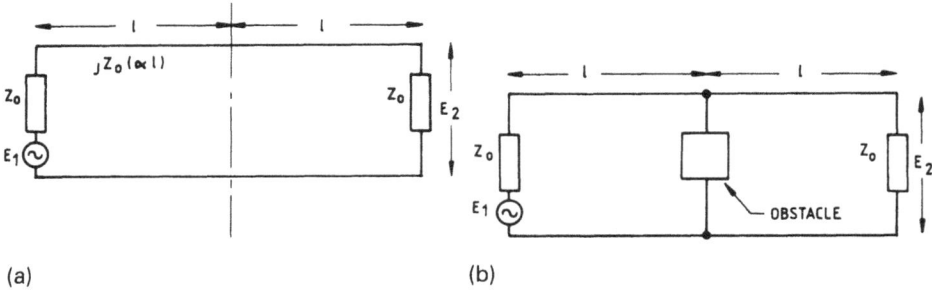

Fig. 3.1   Insertion loss of (a) the cutoff guide section and (b) the cutoff guide section with obstacle.

measured, dissipative losses can be expected to interfere with accuracy. Inductive susceptances should introduce few problems, the only choice of importance being the length l which should be such that, when $-jB \rightarrow 0$, the attenuation should not be too high. The curve in Fig. 9.3 illustrates a typical result. Practical problems in carrying out the measurement involve the manufacture of suitable holders for the obstacle, especially when it is cylindrical.

9.4   Poles and zeros

Frequency is a parameter which can be measured with great accuracy and it is natural to consider its possible application to measurement of susceptance. If, for the moment, we limit the discussion to coupled resonators, the equivalent circuits shown in Fig. 9.4 describe the two coupled circuits. If we assume that the value of the inverter is unknown, then the elements in the network will be uniquely defined if we measure the poles and zeros of the input impedance. By setting up the simultaneous equations which apply to such a circuit we shall be able to determine the corresponding element values. If the two circuits have been tuned to resonance at a particular frequency, the capacitive and inductive values in Fig. 9.4(a) are assumed to be known and the determination of the inverter impedance is quite simple. This is a simple method of checking the inverter value which is derived directly from the length and attenuation constant of the guide at that frequency. When a susceptance is located in the inverter circuit, this technique provides a means of finding its value.

From Section 4.2 we have the equation

$$\frac{X_a}{Z_0} = \sinh(\alpha l) + B \sinh^2\left(\frac{\alpha l}{2}\right) \tag{9.7}$$

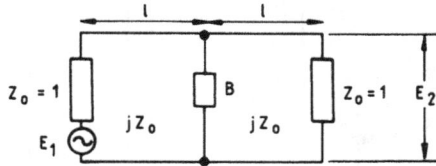

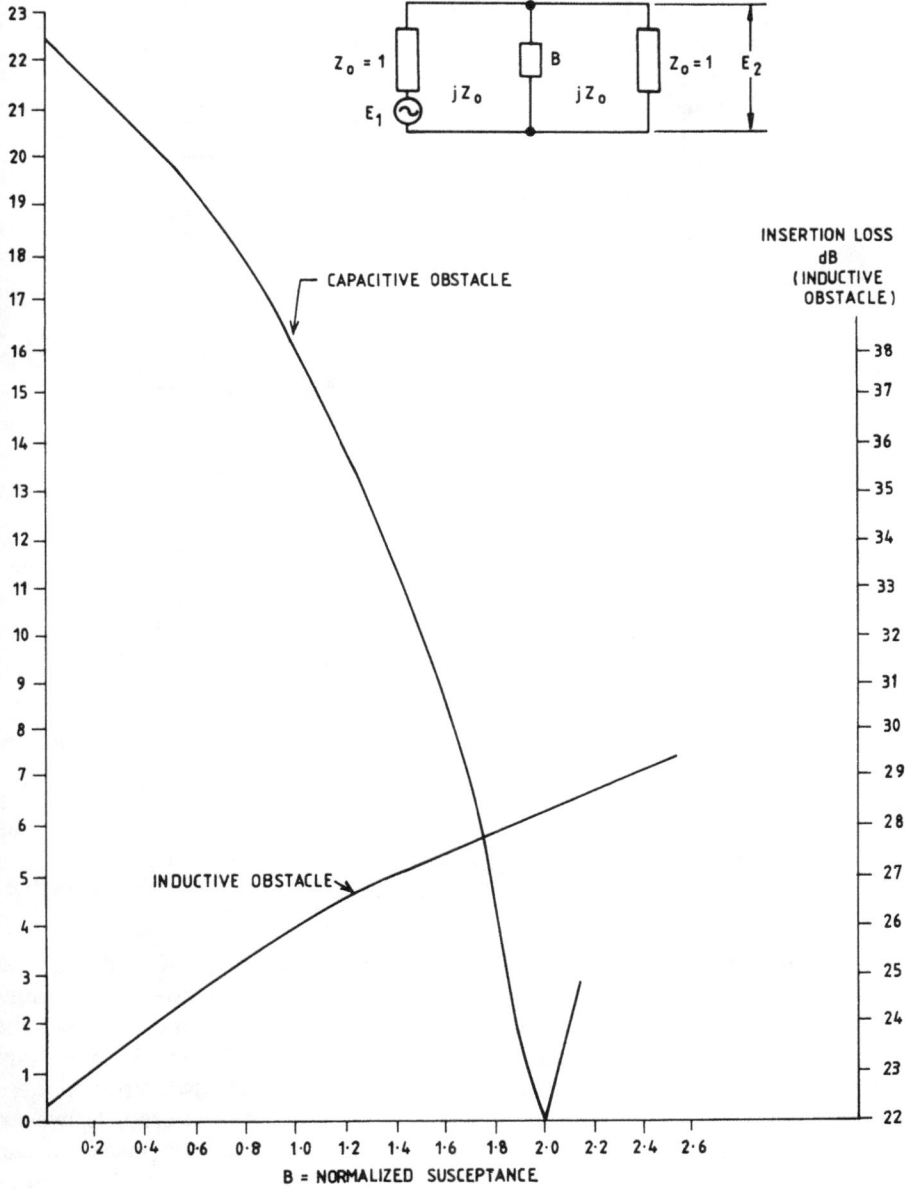

Fig. 9.3 Insertion loss of obstacles: $f = 4$ GHz; $f_c = 6.56$ GHz; $l = 1.5$ cm.

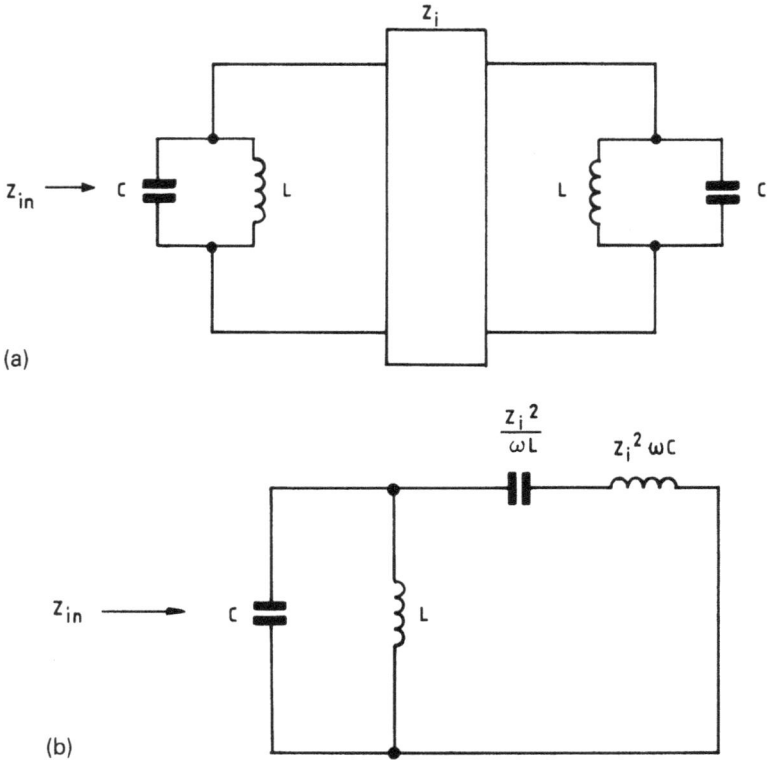

Fig. 9.4 Measurement of Z_i.

or

$$B = \frac{X_a/Z_0 - \sinh(\alpha l)}{\sinh^2(\alpha l/2)} \tag{9.8}$$

It should be noted that it is the value of X_a (i.e. the series element in Fig. 4.2) which is important in (9.8) in its relationship to the magnitude of the shunt susceptance B, since it is the series term in the π-section which determines the inverter magnitude. If the network was wide band (short l), it might be necessary to consider the modified shunt elements of the π-section also, since if these elements change the overall attenuation of the network is altered. However, since we are using the technique purely for measurement purposes we can always choose a length which leaves the shunt elements unchanged. This simplifies the calculation. The foregoing measurement technique was employed in establishing the inductive diaphragms that were used in the elliptic function filter described in Chapter 4.

9.5 Complete networks

The method described in the previous section, with its reference to poles and zeros of input impedance, is a reminder of the basic property of pure reactance networks: their transmission properties are completely specified by the frequencies at which their open-circuit and short-circuit poles and zeros occur. Thus, in the circumstances where a network is not performing properly, we have a potential method for determining the elements in the network that are in error if we measure these critical frequencies. Although in the general meaning of the word this is a problem in analysis, in the more specific sense it is a problem in network synthesis, i.e. we must synthesize the network which is producing this (undesired) response in order to determine which elements are incorrect.

The method is more easily illustrated using an example. We construct a filter (Fig. 9.5(a)) and owing to a design error the response characteristic in Fig. 9.5(b) is obtained; the filter is assumed to be exactly tuned by the method first described by Dishal [3] using poles and zeros of input admittance. If the performance is assumed to be the result of an arithmetical error in the inverter design, a symmetrical network results. Since ordinary microwave bandpass filters are symmetrical this is the more likely result. If we then disconnect the load, the network of Fig. 9.5(a) is a pure reactance and is completely specified by its open-circuit poles and zeros at one input port; these critical frequencies are real (in the sense of being measurable) and not complex, as would occur if the load were retained. Having placed an open-circuit on the filter we measure (calculate using a suitable computer program in the present exercise) the open-circuit poles and zeros of the input impedance. These are shown in Table 9.1 and should be compared with the corresponding frequencies of the correctly designed network in Table 9.2. A final measurement of the admittance at a non-critical frequency will enable us to determine the constant H which then completely specifies the network. The problem is simplified, of course, by the fact that we know the kind of network that we are investigating — a ladder network — and this knowledge will indicate the appropriate mathematical procedures. We then have the choice of two methods.

(a) Obtain the values of the network representing the response characteristic in Fig. 9.5(b) from which we can determine the J-inverter magnitudes in Fig. 9.5(a).

(b) Employ the bandpass-to-lowpass transformation to convert the critical frequencies of Table 9.2 to those of the corresponding lowpass ladder network prototype.

We choose method (b), because it allows us to check the value of the lowpass prototype which may include a wrong value in one of the elements, and we obtain the poles and zeros of Table 9.3. We now have two further options.

$$z_{i1} = z_{i4} \; ; \; z_{i2} = z_{i3}$$

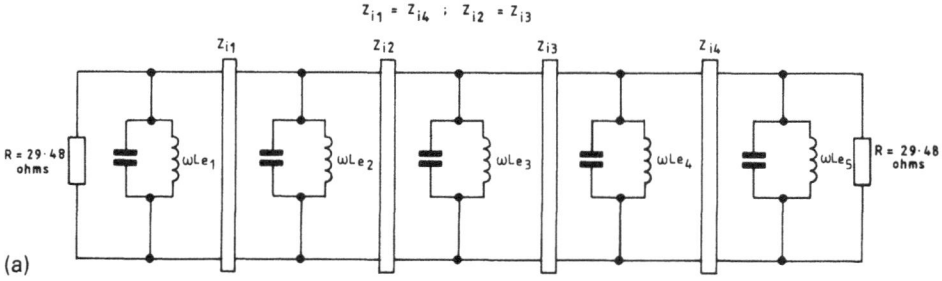

(a)

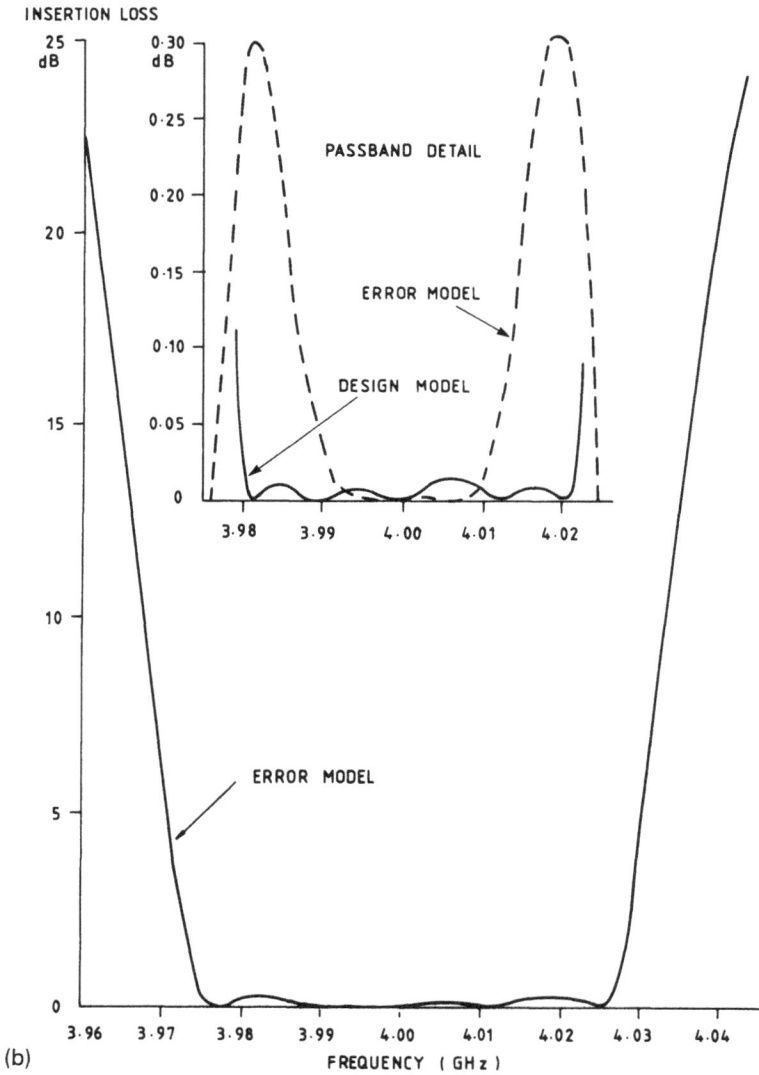

(b)

Fig. 9.5   Inverter-coupled filter: (a) construction; (b) performance.

Table 9.1   Error model

*F (GHz)	3.970311	3.980148	4.00009	4.02043945	4.03078288
○F (GHz)	3.972532	3.988761	4.0117226	4.02847023	

X(at 4.005 GHz) $= -38.1324433$

*F, reactance pole; ○F, reactance zero.

Table 9.2   Design model

*F (GHz)	3.9722643	3.9801291	4.00006411	4.02041958	4.02866312
○F (GHz)	3.9743660	3.9894211	4.01099502	4.02647244	

X(at 4.005 GHz) $= -32.3581364$

*F, reactance pole; ○F, reactance zero.

Table 9.3   Lowpass model (design)

*ω	0.539019158	1.30171052	
○ ω	0	1.00661737	1.40889748

$H = 0.7563$

*ω, susceptance pole; ○ ω, susceptance zero.

(1) Compute the values of the lowpass elements which lead to the corresponding lowpass characteristic by solving the simultaneous equations which describe the various poles and zeros of the network.

(2) Determine the element values by synthetic division.

We choose method (2) because of its direct association with ladder networks, and first we employ the values in Table 9.3 to obtain the value of the constant H. Forming the expression for Y in terms of the poles and zeros of admittance we have

$$Y = \frac{jH(\omega - 0)(\omega^2 - A^2)(\omega^2 - B^2)}{(\omega^2 - C^2)(\omega^2 - D^2)} \qquad (9.9)$$

from which we obtain

$$H = B_{in}/Y \qquad (9.10)$$

where B_{in} is the computed (measured) admittance of the circuit shown in Fig. 9.5(a) when normalized to its source conductance. Substituting $S = j\omega$ we

Table 9.4   Lowpass model (error model)

	A	B	
$*\omega$	0.580786764	1.41401301	
		C	D
$\circ\ \omega$	0	1.01489751	1.52880099

$H = 0.76674$

$*\omega$, susceptance pole; $\circ\ \omega$, susceptance zero.

multiply out the parentheses in the numerator and denominator of (9.9), which can then be expressed in the following form:

$$Y = \frac{a_5 S^5 + a_3 S^3 + a_0 S}{a_4 S^4 + a_2 S^2} \tag{9.11}$$

Equation (9.11) is an example of the special kind of Hurwitz polynomial in which every other term is missing. Solving this equation by synthetic division [4] we obtain the elements of the lowpass structure (Table 9.5). Comparison of the results with the g values of the lowpass design model (Table 9.5) indicates that these values could not possibly have formed the basis of the filter design. Consequently, we check on the lengths which are involved in the filter design. These can be obtained from the expansion of the admittance expression in conjunction with the lowpass-to-bandpass transformation. This well-known relationship can be re-stated as follows:

$$\omega = \frac{f_0}{f_2 - f_1}\left(\frac{f}{f_0} - \frac{f_0}{f}\right) \tag{9.12}$$

where f_0 is the center frequency, f_1 and f_2 are the limit frequencies, f is the bandpass frequency variable and ω is the lowpass angular frequency.

The continued fraction expansion for the admittance of the lowpass network can be written as

$$Y = Y_0 + \cfrac{1}{Z_{i1}^2 Y_1 + \cfrac{1}{\cfrac{Z_{i2}^2 Y_2}{Z_{i1}^2} + \cfrac{1}{\cfrac{Z_{i1}^2 Z_{i3}^2 Y_3}{Z_{i2}^2} + \cdots}}} \tag{9.13}$$

Table 9.5 Computed lowpass values (g values)

Design model	0.7563	1.3049	1.5773	0.7563
Error model	0.7667	1.2656	1.2061	0.7640

Table 9.6 Computed Z_i and length

Design model				
$\sinh(\alpha l)$	38.322	55.344	55.344	38.322
l (cm)	3.985	4.322	4.322	3.985
Error model				
$\sinh(\alpha l)$	38.005	47.666	47.774	38.023
l (cm)	3.977	4.185	4.187	3.985
Actual error values	3.985	4.2	4.2	4.2

By combining eqns (9.12) and (9.13) and solving as discussed in Section 3.2, we obtain the general equation

$$\sinh(\alpha l) = \frac{\omega_0}{\omega_2 - \omega_1} \left(g_i g_{i+1} \omega_0 L_{0,i} \omega_0 L_{0,i+1} \right)^{1/2} \tag{9.14}$$

By performing this operation on the values in the second row of Table 9.5 we obtain the magnitudes of $\sinh(\alpha l)$ given for the error model in Table 9.6. The corresponding values given for the design model in this table were determined in the same way and agree exactly with those in the prototype tables. Of course, the reason for the less exact results obtained for the error model lies in the less precise determination of the critical frequencies which simulate less accurate measurements that might occur in practice. Even so, the actual values used in the experiment and quoted in the last row of the table make it clear that an error in excess of 1 mm in the length separating the second and third (and third and fourth) resonators exists.

9.5.1 Measurement techniques

Some of the errors in the critical frequencies were of the order of 200 kHz. Considerably higher accuracy should be achievable with the method described in this section.

The scheme is illustrated in Fig. 9.6 and is fundamentally a microwave bridge. The filter is mounted on one of the symmetrical arms of a hybrid-T which has previously been balanced for the highest possible directivity. With the short-circuit settings as shown, zeros will be indicated by a minimum of output

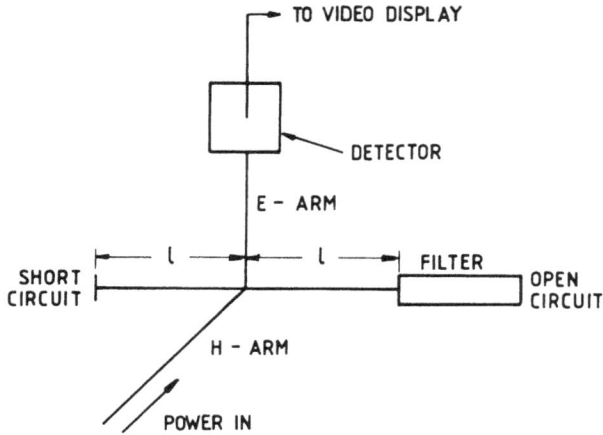

Fig. 9.6 Microwave bridge.

in the E-arm and therefore high gain can be employed in the display apparatus. As a result, if a frequency counter is employed the null frequency — the zero — can be measured to a very high accuracy indeed. The corresponding poles will be represented by maxima and will be less precise, but accuracy can be restored by rebalancing the bridge in order to yield a zero of output at a pole. This requires a quarter-wave shift in the position of the short-circuit. In circumstances where significant loss exists at a particular frequency, it can be advantageous to add an extremely small amount of variable loss in the arm containing the short-circuit in order to obtain a more exact balance of the bridge.

In the foregoing it has been assumed that normal precautions concerning junction susceptances, correct open-circuit conditions on the filter etc. are observed. Further, care should be taken to avoid using poor sweepers, or any other signal sources which have incidental frequency modulation or noise in their output, since this will militate against a good null. If these precautions are observed the critical frequencies should be measured to a high degree of accuracy.

The example was chosen to illustrate complex networks in general. It could equally well have been a subsystem which was not performing correctly. The attraction of the technique is that it provides a systematic procedure for determining errors in the elements comprising a complex subsystem which may otherwise be practically impossible to sort out. It is particularly useful in units of the present type which include no additional lengths such as occur in propagating transmission media. The additional lengths in conventional transmission lines greatly complicate the interpretation of the circuit behavior.

It is reasonable to consider whether the technique can be extended to circuits which terminate in a load. Probably it can, but the possibilities have not yet been investigated. The reflection coefficient K can be expressed in the form

$$K = \frac{j(X_{sc} - B_{oc})}{2 + j(X_{sc} - B_{oc})} \tag{9.15}$$

which defines K in both magnitude and phase. In the above derivation the load is normalized to unity. X_{sc} and B_{oc} are the short-circuit reactance and the open-circuit susceptance respectively, which have to be determined from the real and imaginary parts of eqn (9.15). Once these are obtained the network can be determined by synthetic division as previously described. This approach has the attraction that an ordinary Smith chart representation could be analyzed in this way. Whether the results would be as accurate would require detailed investigation.

References

1 Mok, C. K. Method of obstacle admittance measurement in below cutoff waveguides, *Electron. Lett.*, **6** (3), 43, 5 February 1970.
2 Schieblich, C. and Schünemann, K. Impedance measurements in waveguide below cutoff, *Proc. 11th European Microwave Conf., Amsterdam, 1981,* p. 507.
3 Dishal, M. Alignment and adjustment of synchronously tuned multiple-resonant-circuit filter, *Proc. IRE,* **39**, 1448, November 1951.
4 Van Valkenburg, M. E. *Introduction to Modern Network Synthesis*, p. 92, Wiley, New York, 1960.

Index

163

www.ingramcontent.com/pod-product-compliance
Lightning Source LLC
Chambersburg PA
CBHW021431180326
41458CB00001B/228